In the Public Interest

People, Politics and Power in Tampa Bay's Water Wars

By

Honey Rand

In the Public Interest: People, Politics and Power in Tampa Bay's Water Wars

By Honey Rand

This book first published 2023

Ethics International Press Ltd, UK

British Library Cataloguing in Publication Data

A catalogue record for this book is available from the British Library

Print Book ISBN: 978-1-80441-326-5

eBook ISBN: 978-1-80441-327-2

Dedicated to the people who serve the public interest every day.

Table of Contents

Table of Figures

Acknowledgments

There is no way to thank the people who have been generous with their time and insights for this work. Still, I thank them and hope they believe the outcome is worthwhile.

Without my husband, Jim Randel, this book would not exist. I'm thankful for the encouragement and the occasional kick in the pants.

Preface

Coming Home

Though Boston is among the earliest places settled in the New World, it has maintained strong links to its history. Indeed, it has a booming tourism industry based on visits to old churches and buildings with historical walks and tours. Then there are the parks and green spaces. The green spaces allow locals and visitors to cool off on hot days, enjoy outdoor sports, or lie in the sun daydreaming they are on a Gulf Coast beach in sunny Florida. But Florida hasn't preserved its history nearly as well.

When I was younger, I wanted the old buildings in Sarasota torn down. I loved the new, tall buildings. They made me feel excited about an unlimited potential future. Even so, I left home at seventeen—finishing high school early—and jumpstarted ten years of adventure. To me, Sarasota was a nowhere, nonhappening place. There were two kinds of people; the wealthy who lived at the beach and the those who served them.

My family didn't live at the beach. So I left.

Thirteen years later, after looking all over the world for a better place, I came home. As I sat in the back of the real estate agent's car shopping for a home, I couldn't believe the changes—so many people. The tomato fields east of I-75, where we used to visit "U-Pick-Em" fields, were gone, replaced by one- to five-acre ranchettes and golf courses. "Where," I asked the real estate agent, "did all these people come from, and what do they do for a living?" I couldn't imagine. I came home for a month each year spending my time at the beach. I never noticed the changes. Moving home, I wanted a job that would let me help protect Florida, or at least my little corner of it.

Home at Mote

Mote Marine Laboratory is a marine and environmental research and education center in Sarasota. As a kid, if you found something interesting in the water, you'd call a Mote scientist, who might drive out and pick up the thing. Now Mote is internationally recognized for its research into sharks and cancer, marine mammals and red tide, fisheries, coral, and turtles. The lab boasts Dr. Sylvia Earle and Dr. Eugenie Clarke among its former executive directors. Dr. Earle is a former chair of the National Oceanic and Atmospheric Administration; Dr. Clarke is among the most respected shark researchers in the world. Among water environmentalists, these are well-known names.

I loved the work at Mote. More than just professionally and intellectually challenging, it gave me exactly what I wanted, the sense that I was protecting my home in my way. It researched everything from sharks' cancer resilience to the causes of red tide, to sex-change operations for grouper. I loved it; I was a believer, a zealot. If anything, the hardest part for me was that Mote didn't take positions on issues. Mote would conduct research and report that research, but it would not address the policy implications of it. In one respect, this adheres to the pure-science model that is occasionally invoked in policy debates. It's also a way of sidestepping dicey politics, which can affect future funding. On occasion, I wanted Mote to take a stand. I knew that scientists could say things to influence people. I knew Mote could bring credibility to an argument. Then again, that credibility is not to be squandered. Mote never did.

I spent three years at Mote. We got lots of great news coverage on the research. I managed a national campaign, the Year of the Gulf of Mexico, which brought money into operations and put us in newspapers and publications around the country. I had a knack for translating science into ideas that made sense to people, and the job gave me what I wanted: an opportunity to use my communication skills to help people understand that protecting Florida is everyone's job.

At Mote's open house one year, I met the communication director for the Southwest Florida Water Management District. He told me that he was leaving his job and that the position might interest me—"if you can stand the Rottweilers," he said, exchanging knowing glances with another colleague we knew. "Who are they?" I asked. "Gene Schiller and Mark Farrell," he said, with apparent disdain. The names meant nothing to me.

In May 1993, I was preparing for the Mississippi River Project, a part of the Year of the Gulf of Mexico designed to help educate people about the pollution in the Mississippi River contributing to a dead zone in the Gulf of Mexico. Vice President Al Gore had promised to attend a water-quality testing event in Memphis. We were taking ten thousand students in fifteen states out the same day at roughly the same time to test water quality. We were going to create a satellite link from the students at the Mississippi's headwaters to the outfall in New Orleans. A satellite link in those days was expensive. I would be in Memphis with the vice president. I had a lot on my mind.

It was one week before the event when the water-management district posted its advertisement for a director of communication. I didn't even see it—but Jim Randel, my husband, did. A week before a national event with the vice president, the last thing I wanted to think about was my résumé. Applying meant a cover letter, portfolio preparation, and comparison of my skills to their needs. There was no time. Still, Jim persisted.

"The job will be closed by the time you get back from Memphis," he claimed.

"Will you get off my back, if I do?" I asked.

"I'll even do your résumé if you'll tell me what to do," he said.

So we prepared, and he applied. That was April.

In October, a call came from Eugene Schiller—one of the Rottweilers. We talked. I liked him.

"So what brought my résumé to the top of the pile on your desk?" I asked. Gene gave me some benign response that made me think I might not be able to easily rock his boat.

I always liked a challenge.

More than 750 applicants sought the job. Two rounds of interviews had been done, but Gene and the other executives were waiting for someone—they didn't know who. They invited me to District headquarters to talk.

The Interview at the District

It was November when I drove north to Brooksville. I was escorted upstairs, where I waited a few minutes before being taken to a small conference room. Gene Schiller stood to greet me as I came in. A blonde woman was seated at the conference table; she introduced herself as Linda Sullivan, director of boards and executive support. A tall guy stood at the head of the table. Peter Hubbell was the executive director. Another man remained seated, though we shook hands it was Mark Farrell was the other Rottweiler. Gene gave a preliminary introduction and explained the procedure. This was a panel interview. There were set questions, but each participant was able to ask follow-up questions at will. Then they began. Mark Farrell leaned his chair back against the wall, tipped on two legs. He couldn't be less interested. He was the one I needed to engage.

I remember two specific things about the interview. First, it was clear they needed someone to fix their communications department. Many of the questions focused on staff relations, structuring the work, and creating a management culture. I got the best reaction when I shared that I had previously fired a federal employee. It's an onerous process. Clearly, that's what they needed. Second, I made them laugh. In response to a question about whether I would be able to go into the executive director's office and pound on his desk until I got his attention, I told a story about the death of Buster the Jewfish (the jewfish has been renamed the goliath grouper) at Mote:

> Buster the Jewfish was a four-hundred-pound grouper who lived in the public aquarium at Mote. One morning I came in and was informed by the volunteers that Buster had died that night. When I went to investigate, I found that samples had not been taken for the scientists. They were upset that Buster had been hauled out into the Gulf and unceremoniously dumped by the lab boat captain, who didn't want anyone to mention anything about Buster to anyone.

> "People will notice that a four-hundred-pound fish is missing," I said. "People will think the worst if we don't tell them the truth." We argued, but he was determined. So was I. We could not allow Buster's demise to remain a mystery. The scientists needed their samples, and I needed a cause of death. I appealed to the executive director, who eventually relented. While I was in his office, a call came in. Buster's remains had floated to the surface of the Gulf and were reported heading inshore. The captain went out to recover Buster, but instead of bringing him back to the lab for tests, he took some samples and shot the fish full of holes, hoping to fill the carcass with water and sink it. It was too late. When he got back, we continued our argument about the disposal of Buster. But later that afternoon, we got a call from the manager of a condominium complex on Longboat Key. A massive fish had washed up on the beach, and it had gunshot wounds! They wanted a Mote scientist to come right away and investigate. ("Do you think it was suicide?" I asked, sarcastically.) It was as if Buster wasn't happy with the way he'd been handled and kept coming back until we could get it right. Someone took a backhoe and buried Buster on the beach, where he rests today (we hope) in peace. . .

I intended to be funny, and the illustration was humorous, but it told them I would be willing to "pound on someone's desk" to get their attention if I needed to.

I had them howling. Even Mark. I showed them my portfolio and waited until the last question: "If you were selected for this position, when could you start?" "I've already resigned from my job at Mote," I told them. I felt that it was time to find a more complex assignment. Mote was very busy, and a lot of fun, but the complexity resided in science, not in the ways it was communicated. I wanted more responsibility.

Within days Gene Schiller called and offered me the job.

Home at the District

I started to work for the District before my official start day. I attended the Governing Board meeting in November 1993. Gene Schiller had already told me the Governing Board wanted to "do something about communications." There was a drought, the District been pounded in the media for several years over its failure to provide leadership and take stronger stands to protect the environment, and the board wanted a new direction in communication. Part of that included the allocation of $300,000 for a public education program called Do Your Part, based on research that demonstrated that people were willing to conserve water during low rainfall but only if their neighbors did. The Do Your Part campaign was based on behavior modeling. We would show people how others "just like them" had saved water. The program had great examples of homeowners, agriculture, industry, and business. We used real people with real stories. I attended the November meeting of the Finance Committee, which funded the program. Executives wanted me to go in and explain the program to the Finance Committee. To sell it.

At the meeting, the board members, clearly annoyed with the public-communications staff, turned to me and began to question the program. I'd been sitting in a chair against the wall. Not knowing any better, I pulled my chair up to the table where the board sat and began to explain the broad concepts and the specifics of the program. It was my first test. We got the money for the program. As we walked back to my office, staff and consultants were elated. "Finally," said one. "Now let's get going." Things were about to get going, but none of us had any idea just how much.

Chapter 1
They Could See It Coming

Steve Monsees stood before the Governing Board of the Southwest Florida Water Management District. He was nervous and angry. The hundred-acre lake in front of his house was nearly dry, going the way of the pond behind his house. Steve had seen drought before, but he was convinced that it was not the result of drought. It was something else. He went to the Governing Board to ask for help. He gripped the sides of the podium as he faced them. They looked back impassively. Many citizens had addressed the board before. The board had heard their stories, and they would hear his. But this time was different. Monsees, a retired Green Beret, had decided that this time they would hear him, or they would be very, very sorry.

Water Isn't Everything; It's the Only Thing

No matter how you get news, issues about water dominate, and not just in Florida. There's too much or too little. There are plenty of water-quality issues, too. Some occur naturally, but far more are driven by human activity or failure to plan and invest. That's true for the developed world, but less so in developing countries. And more frequently today there are climate refugees where, again, too much or too little rainfall affects farming and even subsistence living.

In July 2023 the Food and Agricultural Organization urged countries to prepare for water shortages that will affect five billion people across the world. Three billion people living in agricultural areas already experience water scarcity. About 57 percent of the global population will likely face water shortages by 2050 for at least a month every year.

In the US and likely elsewhere, there are the intentional and accidental the incompetent and other causes that cause worry about water quality. Consider Flint, Michigan, where public officials sent contaminated water to a largely minority community and have gone to jail over decisions to go use cheaper, unsafe drinking-water sources. There is Camp Lejeune, where drinking-water supplies were contaminated with industrial solvents. Marines and their families drank that water for decades. There are others—too many others. Alabama, Florida, and Georgia have been in litigation for nearly thirty years. The Rio Grande has fostered conflict between Texas and Mexico. Withdrawals from the Colorado River are dwindling. And while states receiving water have agreed to cuts, except California, smaller users are finding lowered flows may be too low for their intake pipes. The Mississippi River flows are so low that shipping commerce is affected.

Conflict over water isn't limited to the United States. There are conflicts over the Euphrates and Tigris Rivers, the Jordan River, and the Nile. In March 2023, the United Nations urged more countries join the Water Convention to reduce tensions and improve international relations.

In August 2023, the *New York Times* reported that while countries invested in reservoirs, they are filling far more slowly than anticipated. Ever-increasing demand, coupled with changing weather patterns, forces water suppliers to seek new sources, additional conservation, and technology. There is ever-increasing demand and ever-increasing cost.

And in another article, the *New York Times* found that groundwater across the US has declined significantly over the past forty years as more water has been pumped out than nature can replenish.

Florida is the envy of most other states when it comes to water. Florida has, theoretically, abundant rainfall, except when it doesn't. Florida's subtropical climate demands abundant water to maintain ecological health and provide habitat for wildlife. Its beautiful estuaries need freshwater flows to maintain their health. Wetlands hold the water generated from hurricanes and drenching summer rains, but wetlands are threatened by a recent Supreme Court decision. Wetland protection may now largely lie in local elected officials who have rarely shown deference to decisions good for the long haul. Florida's aquifers, the primary drinking-water source for most people, farms, and businesses, are recharged by rainfall. But like the rest of the world, Florida suffers from periodic and prolonged drought and changing weather patterns.

Twenty-five years ago, the Tampa Bay region was gripped by a terrible drought. The Southwest Florida Water Management District had permitted too much water to be taken by local-government public suppliers. Water wellfields in northwest Hillsborough and central Pasco Counties were worked harder and harder, and in the process nearby lakes and ponds were sucked dry. Wetlands, too. People living near the wellfields complained but were assured by technicians at the District that eventually the system would reach "stasis," meaning the water levels would reset, probably at a lower level, and the unrelenting decline of surface waters would stop; but the system never reached stasis. And that water wasn't serving locals. It was piped to coastal Pinellas County and St. Petersburg, where lawns were green and lush.

St. Petersburg's water permit was in for review at the District. It had been for two years. As the drought worsened and the water pumping continued, environmental damage spread beyond the wellfields, affecting tens of thousands of acres of wetlands. District staff were alarmed and had been studying the situation for years, knowing that if they reduced the allocation of water to local governments, there would be political fallout the region had never seen.

That was the situation when I became the director of communication for the Southwest Florida Water Management District. Public support, according to its latest survey, was less than 25 percent. Less than forty-five days after I arrived at the District, Steve Monsees spoke to the Governing Board, kicking off what became the Tampa Bay water wars.

This book is my account of what happened—the strategies, both legal and communication, the political maneuvering, the unrelenting environmental impacts, the sacrifices, and, ultimately, the resolution. Millions of dollars was spent on litigation that didn't produce one drop of water, then billions of dollars was spent to provide new water supplies and recover the wetlands to the greatest extent possible.

Where Water Comes From

No matter where you live, water comes from a few sources: underground, surface, or technology. The technology category is evolving but includes desalination (removing the salt from water) and exotic ideas, like in Texas, where a university is working on making water from air. Underground water is called groundwater and comes from aquifers, which can be deep or not and can have great quality or not. The quality of water drives the cost of treatment. Surface waters can be rivers or ponds. Surface water generally costs more to treat to drinking-water standards because there are more potential contaminants, but all water is getting more expensive because there is higher demand for a limited resource. There is no new water. Earth has a finite quantity, and until recently, most of it was locked into glaciers, but that is an entirely different issue. Technologies like desalination can let us tap the ocean for water, but the energy to extract salt from it is costly in many ways beyond actual expense. Typically coupled with traditional power plants using carbon-based fuels, desalination can add to the greenhouse gas issue, which is driving climate change. In the last ten years, desperate communities have turned to recycled or purified water: the wastewater plants treat water to near-drinking-water standards and discharge it into natural systems to rehydrate them; or they actually close the water loop by treating the wastewater, then sending it back to the drinking-water system to be repurified and sent to customers. It's perfectly safe but suffers from what the industry calls the yuck factor. The media often call it "toilet to tap." That's a technological solution, and it's being used in California and Texas and other typically water-poor states, though Florida has been testing the concept, too.

The thing is, no matter where our water comes from, there is a cost. Take too much water from the natural system, and the environment suffers. That's what happened in Tampa Bay. Too much of the good, cheap water was taken from underground, during a drought, and it had a devastating environmental and human impact.

Even though the water sources are the same, eastern and western water law are very different. In general, the western US supports private water rights and ownership, while eastern water law generally assumes water is public property. That said, private companies can and do get water permits and develop water for sale (that's where all that bottled water at the grocery comes from). When people think about water wars, they think of the West, but the truth is that everywhere there is competition for a limited resource, there is conflict. Cities compete against each other; agriculture competes with cities, golf courses, industry, and recreation, all of which need water. So does the environment. These competing needs create conflict.

The Development of Florida

For most of its history, Florida was considered uninhabitable. Early journals of adventurers, naturalists, and military professionals provide a rich testament to Florida's harsh environment. The central part of the state was considered worthless, full of swamps, mosquitoes, and the "most disagreeable creatures." Nelson Blake's book, *Land into Water, Water into Land*, details how development in Florida is linked to water. He points out that to many of the early visitors to Florida, it possessed a dangerous superfluity—too much water, too much heat, too many strange creatures. There were huge snakes and "such clouds of mosquitoes that travelers sometimes feared for their lives. Even more terrifying were the alligators and crocodiles."

Setting the pattern for Florida's development following the Civil War, Congress established financial mechanisms for ditches, dams, and dikes to control and channel the water, to increase cross-state navigation, and to make Florida more habitable. Florida's overabundance of water had to be controlled. It would be incomprehensible to Florida's early conquerors that a problem of too little water could ever exist here. The earliest settlers and visitors viewed this semitropical paradise in terms of its water abundance. Indeed, more often than not, the problem in Florida has been too much water, not too little. To their thinking, less water meant more land, more land meant more development, more development equaled more growth, and that meant increased prosperity and a better quality of life overall. In *Some Kind of Paradise* author Mark Derr notes that few people at the time understood how Florida's wetness created its exotic appearance and supported its rich diversity of plant and animal life.

While Florida was home to conquering Spanish explorers for hundreds of years and to Native American tribes before that, European settlers from northern climates did not particularly value the state until Reconstruction, after the Civil War. First, doctors recommended Florida's warm climate and warm mineral springs for medical reasons. But once guests arrived they discovered mild winter climates and balmy beaches. That created new and desirable vacation destinations. Florida's abundant and exotic wildlife provided exciting new material for the nation's growing interest in nature and for the naturalists who documented it and the industrialists who exploited it.

In the last century, Florida saw its water resources as a nuisance to be controlled, managed, and diverted. Unlike other states, Florida's water resources appeared abundant because of hurricanes, storms, and natural features like the Everglades, a vast wetland. Consequently, early efforts at water management in Florida consisted primarily of dredging and draining anything wet.

Florida became a US possession through a treaty with Spain in 1822. When Florida became a state in 1845, the federal government owned most of the land. To encourage growth, it deeded land back to the state. Federal lands granted to the state were placed under the control of the Internal Improvement Fund (IIF), a state agency responsible for development. The key to future growth was transportation. While railways were slowly making their way down the Eastern Seaboard, Savannah, Georgia, was considered one of the last encampments of civilization. Jacksonville enjoyed a solid reputation as a location with amenities, but the balance of Florida's coastal settlements remained isolated enclaves of lush tropical splendor. For real development to occur, the railways needed to reach down the coast to make way for supplies, goods, and people, according to Blake.

However, Florida's harsh terrain, wet landscapes, and severe tropical weather, along with the resistance of native Indian populations, made laying track a costly and questionable investment. To encourage the effort, the IIF sold large parcels of land at bargain rates to railroad investors.

On top of the cheap land, the IIF added incentives like land grants and options on future land purchases if agreed-upon goals were met. In addition to encouraging the railroads, this tactic made it possible for the IIF to raise funds for what some believed was its primary goal: draining the swamps. It is not an overstatement to suggest that many saw "improving" the state—making it suitable for new residents and for visitors—required draining the water. Natural history scholars and environmentalists have documented the details of the various plans to drain Florida's swamps to create more fertile farmland and land for development.

The impact on Florida's natural systems and the cost of restoring them are now and will continue to be significant to Floridians and the rest of the country. From well before this century till now, water and its control have been central to state politics. The fights over water have been variously ugly and ridiculous.

A 1944 report from Florida's first water commission signaled the initial statewide interest in water resources. The citizens' committee was to study Florida's "freshwater situation"; the commission was also asked to propose legislation for corrective action.

While that legislation failed, a Water Survey and Research Division was established within the State Board of Conservation. A work of the Florida Water Resources Study Commission was published in 1956 and began with an inventory of water problems. Those problems were recognized before the unprecedented growth of the 1960s, 1970s, and 1980s and well into the real estate bubble of the twenty-first century. It was a time before real urban sprawl, before many endangered and threatened species were identified, before the ecology movement, and before demand for water resources outstripped the available supply.

Of equal importance in the 1956 report is a section on Florida's water laws. Florida subscribes to eastern-US water laws, which hold that water is a public resource belonging to the citizens. The state holds water in trust for the benefit of the people. This system is substantially different from western-US water law, which assures landowner rights to the water on, under, or near their land—based on historical usage. Despite public ownership, the battles over water in Florida are no less contentious than those in California. The first recommendation in the 1956 report was that "a comprehensive water law be established in Florida." Other key recommendations of the report dealt with their revision, supervision, and implementation.

In 1964, Florida's Water Resources Act established funding for a comprehensive study of Florida's water laws. Frank Maloney and Sheldon Plager prepared the resulting book. Plager had once been chairman of the Water Law Committee for the Florida Water Resources Study Commission. He later became its general counsel. Plager was a research assistant to the commission. Their 1968 book, *Water Law and Administration: The Florida Experience,* is a unique document. The conclusion to their detailed effort is eerily prescient: "Failure to undertake such planning now will lead to hit-or-miss solutions which may well result in the freezing of water use patterns in uneconomic and poorly planned projects. . . . Now is the time for Florida to chart her course and provide the legal framework for a sound water-resources program looking toward maximum beneficial use of Florida's most precious natural resource."

By the 1950s, Florida's temperate climate had made it a popular winter home for visitors the world over. Florida's east coast was already well developed by the time Florida's west coast was discovered. In the 1960s and 1970s, it seemed as if everyone wanted to live in Florida. Water-supply and development policies for Florida were not very different from those of other states. But Florida's appealing coasts, its position as a gateway to Mexico, South America, and the Caribbean, its low taxes, and the cooperation of government made it a prime location for development.

The idea of protecting natural environmental systems is a relatively new one. The first Earth Day was held in 1970—more than fifty years ago—just as Florida's growth exploded. Few people really understood how growth would impact natural systems. Fewer still suspected that Florida might have insufficient water resources to provide for its exploding population without suffering severe environmental consequences. Most people believed the water supply was inexhaustible.

But booming populations in West Central Florida have quintupled in the intervening years. According to the US Census Bureau, the Tampa Bay area led the way in 2017 as one of three metro areas in Florida that were among the nation's biggest gainers in the number of people moving there the previous year. The Tampa area had the nation's fourth-highest gain from people moving there in 2016—some fifty-eight thousand new residents. All of them need water.

This growth is reflected in Florida's water history, resulting in conflicts and environmental trade-offs. It is there today.

While various environmental issues exist around the state, this is the story of water-supply policy in the Tampa Bay area. It includes Hillsborough, Pasco, and Pinellas Counties and the cities of St. Petersburg, Tampa, and New Port Richey.

Pinellas County, like Florida, is a peninsula, surrounded on three sides by water. The rapid growth along the beach and bay, which rimmed Pinellas and its coastal cities, is mirrored in Florida's coastal growth. Indeed, Pinellas is the most densely populated community in Florida.

Communities in Southeast Florida had huge water-transfer and water-control structures in place before anyone understood the potential consequences of such policies. But the Tampa Bay development boom came at a time when technology was improving and an environmental ethic was growing among the public. The sources for drinking water were no longer out in the middle of nowhere; they were in the midst of desirable, growing, and largely rural communities.

Tampa Bay's Water History

This water war wasn't the first. For decades, so-called water wars plagued Tampa Bay. Some say the roots of the conflict go as far back as the 1920s, when Hillsborough County extended to the Gulf. Coastal leaders stripped the peninsula out of Hillsborough and created a new county—Pinellas. Others argue that increased demand for a limited water resource created the same problem for the communities of Tampa Bay that it had for growing communities around the globe, wherever demand for water exceeds a cheap, available resource. Whatever the reasons, the legal and political battles over water supply in Tampa Bay have raged for decades, with a legacy of suspicion, competition, and negative environmental consequences.

As the population along Florida's west coast exploded, so did the demand for water. Between 1970 and 1990, the coastal population in the Tampa Bay area grew nearly 48 percent. Between 1960 and 2017, the population of Florida grew from just under five million residents to almost twenty-one million. Much of coastal Pinellas County, and particularly St. Petersburg, outgrew its water supplies, which were overpumped and spoiled with saltwater intrusion. According to the Water Resources Atlas of Florida, saltwater intrusion is a problem for Florida's coastal communities. Freshwater is surrounded by saltwater, and saltwater is heavier than fresh. Lowering the freshwater levels allows saltwater to seep in, reducing water quality to the point that it becomes undrinkable. Today, sea-level rise is making the situation that much worse and that much faster. Already, some coastal communities have had to abandon public-supply water wells, the only alternative being to desalinate the water. More expense.

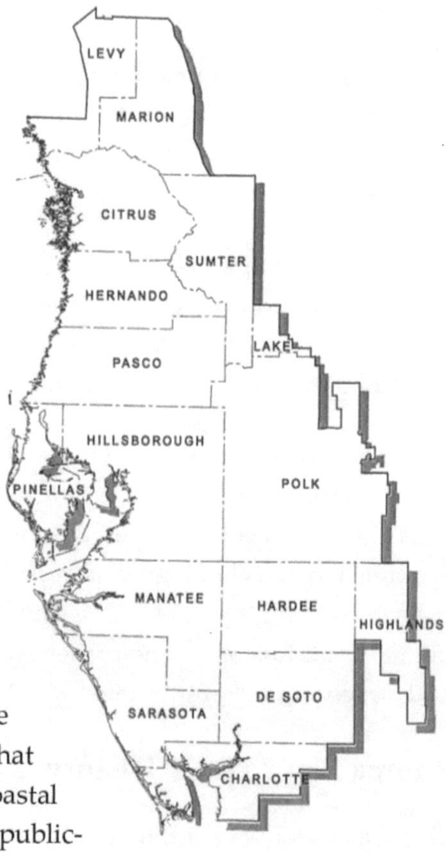

To provide water for their residents, Pinellas and St. Petersburg developed water wellfields in sparsely populated Northwest Hillsborough and Central Pasco Counties, located inland, East and North. But as those areas developed, residents didn't want the big water wellfields in their communities. In some cases, the wellfields were there first; in others, wellfields were installed near established neighborhoods, ranches, and farms. As wellfield pumping increased to meet growing water demand, residents living near the wellfields complained of dropping lake levels and associated impacts they claimed were caused by the wellfields. From the early 1970s until the mid-1990s, their complaints were largely ignored or refuted by government agencies.

In the 1970s, Hillsborough and Pinellas had a bitter feud over where drinking water would come from and where it would go. Pasco County, still quite rural at the time, was only marginally involved. Lawyers were retained, and the political bickering that had festered for decades escalated into the first round of what area newspapers called the Water Wars. As conflict over water escalated, political solutions were sought. Two organizations were central to the resolution of the conflict: the Southwest Florida Water Management District (the water regulator) and the West Coast Regional Water Supply Authority (the water-supply wholesaler). Both were established by an act of the Florida legislature.

The two organizations are respectively referred to as the District and the Authority. In the narrative that follows, the District is sometimes called Swiftmud, which is the pronunciation of its acronym (SWFWMD).

The Seeds of Conflict

In an era in which regions compete for clean industries, sporting events, and tourists, the infighting over water resources in the Tampa Bay region had serious economic consequences. The interdependence of community relationships seemingly should strengthen ties and facilitate communication. Conflict over a primary resource—water—threatened to undermine other cooperative economic initiatives. For example, in the 1990s, a Korean microchip factory was considering a new factory site in Florida. Tampa Bay executives worried that headlines over water-supply problems would send the corporation elsewhere in the state. Chip factories require millions of gallons of freshwater for daily production, but they bring high-paying jobs. I was called to meet with a prominent business leader in Pinellas who, as he told me about the chip factory, complained that the headlines were "killing us."

"You know what happens when we can't get water?" he spat at me.

"Yes. Deals go south."

"No," he snarled; "they go elsewhere." The chip factory wasn't located in Tampa Bay.

The Tampa Bay region's water supply was rainfall dependent. When it rained, there was plenty; when it didn't, the environmental consequences were evident to both the old-timers and the new arrivals. People who lived near the water-supply wellfields provided anecdotal information about problems since the first wellfields were established in the 1980s, but they were ignored or told that their evidence didn't fit the scientific models shared by the District and the Authority. Both the Authority and the District agreed that it was a lack of rainfall—not wellfield pumping—that caused the environmental issues.

By 1987, the District began a comprehensive inventory of water resources in the most impacted area (the area of highest concentrated pumping and water withdrawals). The study took seven years to complete, but well before it was accepted as final, District scientists had determined that, contrary to their prior conclusions, a lack of rainfall was not the primary factor in the environmental impacts to wetlands and lakes in Northwest Hillsborough and Central Pasco Counties. The problems were caused by wellfield pumping. The District reversed its long-standing policy. The people living near the wellfields rapidly allied with the District, as did concerned citizens and the governments of Pasco and Hillsborough Counties.

A water gauge in an impacted wetland, central Pasco County.

The Authority, St. Petersburg, and Pinellas County continued to insist that any environmental issues were due to ditching, damming, drought, and development. The District rejected this premise, and what had been agreement among governments (that insufficient rainfall caused the problem) became a significant battle over water-supply policy. Local governments, citizens, and special interest groups were all actively involved. The District's change in its position became a catalyst for political upheaval and the realignment of political power. The District responded to political pressure in a manner substantially different from ever before—it ignored it. This was disturbing to the Authority, St. Petersburg, and Pinellas County, which had previously pressured the District into adopting policies favorable to them. Moreover, the District's aggressive communications program in support of its new position further aggravated them.

The conflict and drama surrounding water supply in Tampa Bay is a story of the fight for power, money, and control. It was a fight over existing investments and future economic viability. It is also a story about people who believed they were fighting for the public interest. Everyone involved in this story thought they represented a vital public interest. They disagreed about boundaries and the definition of "the public," and they fundamentally disagreed on where the public interest resided. The story of Tampa Bay's last version of the water wars provides an excellent illustration of how good people with good intentions can fail to reach agreement and how far-reaching the consequences of that failure can be. It is also the story of leadership and vision. This is one perspective on Tampa Bay's water wars. The people, the situations, the challenges, and the choices are like those throughout Florida and Texas, Pennsylvania, Georgia, the Carolinas, and the world. The western US is currently living through what some have called a thousand-year drought. It has less rainfall where it rains and less snowmelt because of climate change. As of 2022, one tiny town in California faced the prospect of running out of water in a few months; another had residents using bottled water for everything.

Water is a limited resource. It's the single most controlling factor for growth. It truly is blue gold. Where there is competition for a finite resource, there will be conflict. That conflict will be played out by people—with the same responsibilities and roles as those in this story.

Chapter 2
The District and the Authority

The Southwest Florida Water Management District

Florida is divided into five water management districts based on hydrological features.

The Southwest Florida Water Management District was created in 1961 to manage the flooding in West Central Florida. For years, the District kept a low profile. With a handful of employees, most of them former Army Corps of Engineers staff, the District acted as a funnel for federal dollars aimed at creating large water-control structures.

One example of those projects is the Tampa Bypass Canal, which is a significant water-control structure designed to divert floodwater around the city of Tampa and communities in the vicinity.

But as Florida and Tampa grew, so did the need for environmental management. The Water Resources Act of 1972 expanded every district's authority and provided the legal grounds for water-resource regulation. By the late 1980s, the districts had become water-resource *regulators*—just as people were recognizing the limits of the water supply and the environmental consequences of developing it. Every new water-supply project has an environmental impact. If projects are designed and managed properly, those impacts can be minimized. If they are not, developing water supplies can have severe negative environmental consequences, as it did in Tampa Bay.

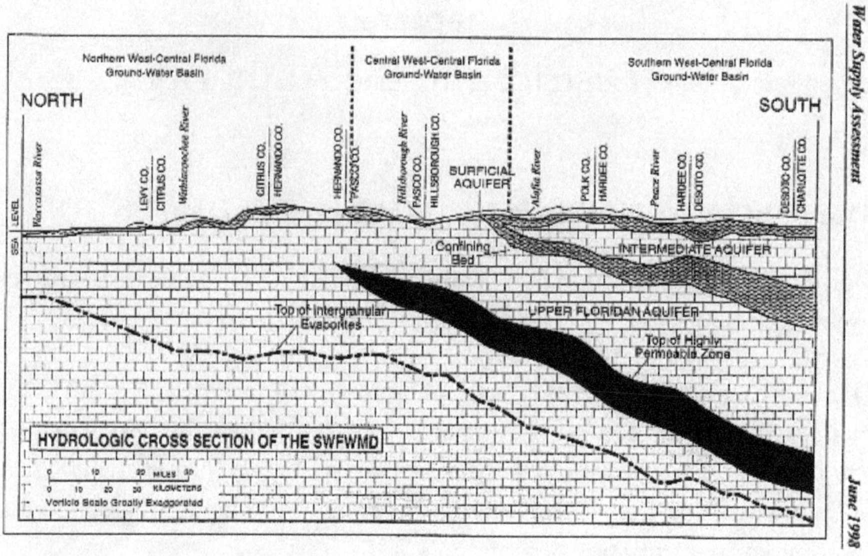

In the 1970s, unprecedented growth created intense competition for water in the Tampa Bay area. The solution to the intergovernmental conflict at the time was the establishment of the West Coast Regional Water Supply Authority in 1975. The Authority was to collectively develop water supplies for the Tampa Bay area. At the time, Pinellas County was the largest and wealthiest water user, making it the most influential in the region and in the Authority. Pinellas County commissioner Chuck Rainey represented Pinellas County on the Authority board, eventually serving thirty years on the Pinellas Commission and the Authority. There was never any doubt who Commissioner Rainey worked for—the people of Pinellas County. He delivered for them, and water was his issue.

To the other local governments, Rainey was a powerful political force to be challenged only when victory was guaranteed. Of course, in politics, there are few guarantees, so Rainey was seldom challenged.

The District was funded by property tax dollars on an ad valorem basis. The Authority was primarily funded by its member governments.

The Authority's purpose was to provide a mechanism for member governments to develop needed water supplies cooperatively. The premise was that a cooperative agency, comprising the significant communities at the time—Hillsborough, Pasco, and Pinellas Counties along with Tampa and St. Petersburg—could more cost-effectively develop needed resources. And it might contain the adversarial issues.

At the time, Pinellas and St. Petersburg were growing rapidly—their need for additional water supplies prompted the siting of new water wellfields in Pasco and Hillsborough Counties, which were mostly rural and agricultural. Tampa had the Hillsborough River as its primary water source. By the late 1960s and early 1970s, Hillsborough and Pasco Counties began objecting that Pinellas and St. Petersburg were "taking water" from them. The Florida legislature believed that the establishment of the Authority in 1975 would put an end to intergovernmental competition and conflict over water-supply development issues.

At the time the Authority was formed, Ed de la Parte was in law school at the University of Florida. His cousin Louis de la Parte had been a legislative leader who established the Authority. When Louis left Tallahassee, a friend of his, Commissioner Chuck Rainey, asked him to become the general counsel of the new organization, the West Coast Regional Water Supply Authority. He agreed.

Ed de la Parte interned for his cousin and worked on water issues for the Authority. The conflicts over water in Tampa Bay have been part of Ed de la Parte's professional life since the beginning. He was intimately familiar with the water conflicts of Tampa Bay, the history of the relationship between the Authority and the District, and the people pushing agendas in public and in the shadows. Whenever there has been disagreement, he has sided with the Authority or its member governments. To the public, the Authority and District seemed mostly on the same page—until they weren't.

According to Ed de la Parte, the Governing Board of the District wanted to act as both regulator and water supplier. He still sees the issue as a conflict of interest.

"How can they regulate themselves?" he asked. "Where are the checks and balances?" Underscoring the differences in perspective, Derrill McAteer, the District's chairman at the time, flatly denied it and pointed to newspaper articles that quoted him on the matter.

"Did we consider providing water? Yes, we did. But we always said it just wouldn't be right for us to regulate water supply and be a water supplier, too. We would be too powerful."

Still, local governments were aggravated by the District's growing responsibilities and the prickly personality of McAteer, who stubbornly resisted their political pressure. McAteer distinctly remembers a meeting with St. Petersburg's state legislator, Betty Easly. "She told me, 'Either you supply the water or give us the chance to have the authority. Your choice.'"

"There's no way the District can be both regulator and supplier, and the public still have confidence in us," he told her. "Absolutely not. Go with your water-supply authority." So the District provided the funds for the establishment of the Authority. The seeds of conflict were sown early between the two agencies and seemed destined to escalate.

Meanwhile, a shaky truce among the local governments, coupled with funding from the District, created the opportunity for new water-supply development—through the new Authority.

The District and the Authority, although similar in their missions, had different priorities. It was and continues to be the job of the District to *protect* water resources while making sure they are available for all users—including the environment. It is the job of the Authority to provide affordable water to its member governments while protecting the environment. It is this conflict in mission priorities that was the genesis for an ongoing, decades-long struggle between the two organizations. The Authority is responsible for providing water supplies for its member governments. The District must concern itself with all water users in all or part of sixteen counties, including users in the fields of agriculture, industry, recreation, and public supply. As the regulator, it is responsible for protecting the environment while ensuring water is available for all users. Given ever-increasing demand and limited resources, it was and remains a formidable challenge. Policy decisions made in Tampa Bay had implications for the entire region.

While the District was established to control flooding, its mandate changed significantly over the years as the population grew and brought attendant problems. The 1992 mission statement said its mission was "to manage the water and water-related resources for the people through regulatory and other programs. Central to the mission is maintaining the balance between the water needs of the current and future users while protecting and maintaining the natural systems which provide the District with its existing and future water supply" (Southwest Florida Water Management District 1998).

Over the years, the District, which has taxing authority of $1 for each $1,000 in property value, was delegated increased authority and responsibility by the Florida Legislature and the Florida Department of Environmental Protection. In everything from water-supply regulation to water-supply development, the twenty years between 1974 and 1994 saw environmental law, resource regulation, technology, and public values change dramatically.

Struggling with these significant changes, the governor-appointed, citizen-volunteer, unpaid Governing Board of the District sought to represent its diverse constituencies while serving the public interest. The idea was that if Governing Board members do not have to worry about reelection, they are insulated from political pressure and, therefore, free to make long-term decisions. From a practical standpoint, it doesn't work that way. Board members have no reelection pressure in the traditional sense of the word election, but they are subject to pressure from the governor, legislators, the public, business interests, and others if they want to be reappointed.

Growing from a small organization with a handful of employees to one with more than 750 staff members in less than thirty-five years, the District grappled with the demands of its responsibility, with the technology to improve decision-making, and with the various powerful interests competing for water supplies. While the District is responsible for all water users, public supply is arguably the most important. And yet the District's relationship with its largest public water supplier seemed persistently contentious.

The West Coast Regional Water Supply Authority

The Authority was established in 1975 as a vehicle for the local governments of Pinellas, Hillsborough, and Pasco Counties, as well as the cities of St. Petersburg and Tampa, to develop new water supplies cooperatively. Many people believe that the creation of the Authority was a political deal cut by powerful Pinellas commissioner Chuck Rainey to ensure that there would be a counterbalance to the District's regulatory authority.

While new water supplies were brought online in the 1980s, they were quickly allocated as growth exploded. Pinellas County, through its belligerent representative on the Authority board, refused to fund the development of new supplies that were not groundwater. As a result, many people presumed that any solution to water-supply problems would not come from the Authority. Pinellas County controlled the Authority. The golden rule prevailed: he who has the gold rules.

In the late 1970s and early 1980s, Pinellas and St. Petersburg continued to develop their water supplies independently. By the time their last water wellfields were in place, the political environment would not tolerate more sources in other communities. Some people felt that too much water was already being taken from the water wellfields, negatively impacting nearby lakes and wetlands. Residents complained about property values and reduced quality of life. Water was exported from Hillsborough and Pasco to Pinellas. Pinellas and St. Petersburg insisted they were entitled to the water from Hillsborough and Pasco. They denied that their water withdrawals caused any problems. Florida's water law allowed the water transfers. A period of multiyear-deficit rainfall in the late 1980s and early 1990s underscored the problem. In the 1980s, the Authority struggled with conflict among its member governments, mostly over the issue of new water-supply development, as Pinellas County stubbornly refused to fund new supply development. The Authority also struggled externally with the

District. Ongoing conflict meant the Authority was stuck. It was forced to work existing sources harder to meet escalating demand.

The Relationship between Permittee and Permitting Agency

The District issued permits for water-supply development (consumptive use) with scheduled, periodic reviews and specific conditions. The permits are permission slips to use the public's water in a limited way and for a limited time with certain conditions. A permit allows for the periodic review of environmental conditions, unintended consequences resulting from the withdrawal (like environmental changes), and potential improvements. When the last public-supply water-use permits were issued in 1988, the Authority was told environmental problems resulted from water withdrawals, and it was unlikely to get the same amount of permits again. Some people claimed not to know that any change in the permits was imminent, while others said they knew. There is no record of a written warning in District files, and local governments insist no such communication ever took place. Still, District leaders insist that good, solid science revealed the environmental consequences of concentrated water withdrawals. Those consequences went beyond harm, which is the standard for denying a permit. "Significant harm" is the standard of protection for minimum flows and levels in water bodies. Florida water law requires that the state water-management districts or the Department of Environmental Protection establish minimum flows and levels for aquifers, surface watercourses, and other surface-water bodies to identify the limit at which further withdrawals would be significantly harmful to the water source or ecology of the area.

According to the District, the last permits were only granted to give local governments sufficient time to develop new supplies to replace those quantities that were to be reduced. But through the 1990s, no new water supplies of consequence were developed. The District was and is responsible for issuing the permits for water supply and monitoring the impacts of those withdrawals. By law, the District could not issue a permit that would harm the environment, and the District knew the permits for public supply, in the already-permitted quantities, harmed the environment. For its part, the Authority and its member governments had made a tremendous investment in their water supplies and transport systems. Cutbacks from existing sources would increase the cost of water for their customers.

Since 1992, when the St. Petersburg permits came under review, District staff had been working to gain cooperative agreement from the Authority to mitigate the environmental problems.

St. Petersburg's permits were in-house for review and had been for some time. The District wanted to be fair, but St. Petersburg initially, on advice from its legal counsel Ed de la Parte, refused to cooperate. St. Petersburg changed legal counsel and its approach to the District, but its permits were still under review, and water-shortage plans—needed to manage drought—were still in development. Then, in January 1994 when Steve Monsees addressed the Governing Board, everything changed.

A Critical Moment

Water lawyer Ed de la Parte says the relationship between the District and the Authority was cyclical, with the relationship sometimes more partnership and sometimes more conflict. While the relationship between the agencies—permitter and permittee—continued its uneasy routine, public attitudes were changing and so was the District. In the mid-1980s, scientists were conducting post-audits of some of the permits issued by their agency, and they were increasingly convinced that too much water was being taken from the natural system. A post-audit is a comparison of the predictors and actual outcomes of permitted withdrawals. Daily, District scientists fielded phone calls from people who believed the consequences of wellfield pumping impacted them. Scientists didn't yet have the proof they needed to stand up in court, but they had an uneasy feeling that something was terribly wrong. As they were developing a plan and educating the Governing Board, they were overcome by events.

Chapter 3
The People, the Policies, and the Possibilities

The fight over water in Tampa Bay is not a new one. It started well before the water management district was established and well before the Authority was created. The relationship between the two agencies is little more than an extension of the litigious fights among the local governments since the early 1900s.

As the region developed and inexpensive water sources became more scarce, competition led to conflict. Pinellas, Hillsborough, Pasco, and St. Petersburg were at odds over the water supply for generations. In those early water wars, science wasn't a factor. Back then, political solutions were used to resolve intergovernmental conflicts. Today, while science is a factor in the decision, the political process still provides a significant means of resolution.

In the 1970s, while the governments around Tampa Bay bickered over groundwater supplies, Tampa had a secure drinking-water source the Hillsborough River. In 1974, when the Authority was formed to stop the fighting, Tampa was a voting member, although Authority votes rarely affected Tampa's water supply.

Still, Tampa's mayor, Dick Greco, generally took a regional perspective on most issues, and the Authority was a regional solution to a regional water-supply problem. It made sense to participate. This was particularly true when the outcome didn't negatively affect Tampa's future.

Tampa's security—with the Hillsborough River as a water source—placed the city in an unusual position. For years, Tampa was the only Authority member that sent a staff member to represent it. Everyone else sent an elected official, a county commissioner, or a city council member. Tampa's representative was the inside/outsider on the Authority board and frequently cast the vote that swayed the Authority in favor of positions taken by St. Petersburg and Pinellas against the positions of Hillsborough and Pasco. The Tampa representative was the utility director, who understood and was sympathetic toward utility water-supply issues.

Greco's longtime relationship with Pinellas County commissioner Chuck Rainey had little influence, if any. A staff engineer, not an elected official, represented Tampa. From an engineering perspective, the positions offered by Pinellas, St. Petersburg, and the Authority made sense. From a political perspective, they did not. There was no fresh water available in Pinellas County. Since Florida law did not prohibit it, St. Pete and Pinellas were within the law to develop the water supplies they needed in other communities. Legally they could do it, technically, they could do it, but politically, it was a growing problem.

While Hillsborough, Pinellas, St. Petersburg, and Pasco argued over where the water would come from, Tampa was booming. Mayor Dick Greco focused on development, and the city was the heart of Tampa Bay—or wanted to be. He managed to sidestep the intergovernmental conflicts over water but knew that to sustain a building boom, he needed to ensure the necessary infrastructure, including a reliable water supply. The Hillsborough River was a great source during average rainfall, but Mayor Greco had seen the consequences of periodic drought and was determined that his city be protected. One particular incident made quite an impression on him. It was a call from St. Joseph's Hospital.

"They called to tell me that the water pressure was so low, they couldn't wash the babies," he told people, still outraged that it could happen in his city. He was determined that it never happen again. "They couldn't wash the babies," he said, shaking his head. "That is just not acceptable. Certainly not for Tampa." So the mayor began developing additional water supplies, including the Morris Bridge wellfield in Hillsborough County. Tampa is located in Hillsborough County. He also pushed for better wastewater treatment with the Hooker's Point Advanced Wastewater Treatment Plant, a project begun during his first tenure as mayor a decade before.

Initially, it was a state-of-the-art wastewater-treatment plant that would help improve the quality of effluent discharged into Tampa Bay. For ten years, the city studied the quality of the water, and in the early 1990s, it was proposed as an alternative drinking-water source. Shocking and well ahead of its time, the idea never gained enough public support to make the project a viable possibility. Additional details on the use of treated wastewater as a drinking source come later in this story. In the 2010–20 time frame, control of reclaimed wastewater became the source of tension between the Authority and Tampa. Though the conflict was largely out of public view, it occasionally spilled into the newspapers, though not with the same intensity as past conflicts.

In the early 1970s, while Tampa was growing, the water-management district was busy trying to inventory the water bodies of Pasco, Hernando, and Citrus Counties. It hoped to characterize their relative health, watersheds, water volumes, wildlife populations, and other information to provide baseline data. The District was also working on big flood-control structures. Before the West Coast Regional Water Supply Authority was created, local governments fought over available water supply while the District concerned itself with preparing for the next big flood.

When Sonny Vergara came home to Brooksville from the Marine Corps in 1971, he wasn't sure what to do with himself. But a chance meeting with the powerful father of a school friend, Jake Varn ("Big Jake"), led him to seek employment at the Southwest Florida Water Management District. Admittedly, he knew nothing about water management. He'd gotten a degree in communications from California State University, Fullerton while in the Marine Corps Reserves. His friend Jake Varn ("Little Jake") was at the District. District executive director Dale Twachtman had spent some time as an Army helicopter pilot, and the two men hit it off. Twachtman gave him a job.

He was at the District for about seven years. When he got the opportunity to work in Tallahassee, he jumped at it. "Not working in the bureaucracy, not just somebody in the organization, but as a special assistant to Big Jake," Sonny said. He wanted to make a difference, to be a big deal. But his time was limited. In late 1979, after less than a year, Vergara was selected to be executive director of the St. Johns River Water Management District. The St. Johns District covers nineteen counties of Northeast Florida, including Jacksonville. Like most of the other water-management districts,

St. Johns regulates resources in heavily urbanized, rural, and pristine places.

Established in the mid-1970s and one of the big three districts, St. Johns had rapidly gone through several executives. "There were enormous responsibilities," Sonny remembered. "They were trying to find their way, and the board was not clear on what they needed to do or how to deal with staff. When I got there, it was a mess." It took him less than five years, but Sonny had his staff build a plan, gain community support, and move the agenda forward.

"It all had an expense," he said. "I drew the ire of a lot of people. I put the St. Johns District through so many changes in such a short time, I gathered too much baggage." That political baggage had its consequences. In several sweeping moves, the chair of the Governing Board, who had been loyal to Vergara, was replaced. "I got calls from a powerful Florida senator telling me that I was next," Vergara said.

After being badgered in a public budget meeting by one contentious board member, the one that had engineered the removal of the chair, he'd had enough. He stood up, dropped the clipboard he was holding, and walked out. He never went back.

Peter Hubbell, a hydrogeologist, joined the Southwest Florida Water Management District in 1978. Derrill McAteer had been chairman for ten contentious years. Sonny was there at the time, too. Hubbell recalls that there were only a hundred or so employees.

"We knew everyone's name, their wife, their kids, even their dog," he joked. "Lots of us were young and single, and we got together and had a good time." There was a volleyball net in the back of the office building. "We'd go out at lunch and play, then on Friday nights, we'd go drink beer," he said. Pete remembers those days clearly. A sociable sort, he works hard but enjoys a good time with friends and colleagues. He is well liked.

It's not difficult to imagine District work in those days. Field data were collected, typically on camping trips that might last for days. The information was brought back to the District, where it was punched into key cards and run through computer systems. "It might take all day, or even a week to do a run," Pete remembered. "Then you'd get to the end of it and realize there was a mistake in a keycard and"—he groaned—"either guess the result if it wasn't too bad or do the run again." The fieldwork was good, the technology was frustrating, and the results were used to inventory resources. Still, the results of the work became part of the District's legacy and the basis of legal challenges to District regulations in the years to come. Some of that work withstood the test of time and is still used as benchmark research.

According to Hubbell, the relationship between the District and the Authority was never a great one. Back in the early 1970s, the District had forced a cutback to St. Petersburg's Section 21 wellfield in northwestern Hillsborough County. "It happened before my time," he said. "But I recall hearing people talk." Governing Board chair Derrill McAteer took a lot of credit for forcing the cutbacks. Although Governing Board positions are unpaid citizen-volunteer positions, they wield considerable power. But unpaid board members can pay personally for their policy decisions. Those unhappy with board decisions sometimes cry, "Taxation without representation." Sometimes there are consequences to business deals.

Even then, twenty years before they took it to court, concentrated (and increasing) wellfield withdrawals had environmental consequences. Pinellas and St. Petersburg wanted unhindered growth, while McAteer—who came from a rural area—wanted to protect the environment. The battles were acrimonious. "We didn't have the science, and we didn't have the rules," McAteer said. "But we were trying to do the right thing. I just didn't think that it was fair for St. Petersburg and Pinellas to do that to Pasco and Hillsborough Counties." For their part, St. Petersburg and Pinellas denied, as they had all along, that pumping could be related to local environmental problems. They insisted that a twelve-year drought, not their water wellfields, caused the problem.

After ongoing, bitter public exchanges, St. Petersburg and Pinellas promised voluntary cutbacks if the District would support the development of the Cypress Creek wellfield in Pasco County. Pinellas and

St. Petersburg had already jointly agreed to fund a huge pipeline from the proposed wellfield. Initial estimates suggested Cypress Creek could provide as much as fifty million gallons of water per day—enough to allow cutbacks and provide for future growth as well. It would be years before the wellfield was developed. By then, scientists demonstrated that Cypress Creek couldn't possibly provide fifty million gallons per day. It might provide half that, maybe less. Following off-again/on-again political maneuvering, the District and Pasco County approved the deal.

Both Pinellas and St. Petersburg resisted the District's authority to regulate water resources. Eventually, St. Petersburg conceded, but Pinellas continued to resist. According to a 1972 *St. Petersburg Times* editorial, "In February, the District advised the Pinellas County Commission that regulation was coming and that the commission should prepare supplemental water resources for overpumped Eldridge-Wilde wellfield. The County Commission did nothing." Pinellas refused to recognize the District's authority and continued to deny environmental impacts.

One thing is certain: by the late 1970s, based on those limited cutbacks, "the lakes and wetlands did respond and recovered a bit, but it didn't help the relationship between the District and St. Petersburg," Hubbell said. "That carried over into the relationship with the Authority." When the cutbacks to St. Petersburg's Section 21 wellfield were ordered, McAteer's motivations were questioned.

"When we cut back Section 21, there was a lot of noise," McAteer admitted. Publicly, McAteer was accused of trying to thwart growth and favor agricultural interests. Growth had exploded, and already the beautiful beaches of Pinellas were drawing thousands of people each month. The joke was people visit twice, then move there to stay. Pinellas and St. Petersburg needed water—they didn't need regulation (interference) from the water-management district.

"They didn't want to be regulated because it would slow down their growth," said McAteer. Pinellas and St. Petersburg officials had agreed to jointly fund an eighty-four-inch pipeline—capable of carrying more than a hundred million gallons per day from the Cypress Creek wellfield in Pasco

County. They were determined to get the water they wanted and determined that nothing would stand in the way.

The District began to understand the technical and interrelated environmental issues associated with wellfield pumping. Scientists started to understand the limits of the natural systems to supply water and the consequences of continuing to take it. But they started with nothing—no rules, no regulations, and limited technology.

"We didn't have the technical knowledge," McAteer remembered. "You try to do your best and come up with some answers. Good scientists give you advice; you take that advice and turn it into public pronouncement."

In the meantime, the political climate was changing. By 1980, McAteer's last year as "chairman for life" as Pinellas commissioner Chuck Rainey dubbed him, the District was twenty years old, it had a much better inventory of resources, and it was delegated more responsibility by the Florida legislature each year. The District was growing. By the early 1980s, there was public demand for increased environmental protection—and correction of past mistakes.

Mark Farrell joined the District in 1984 just as some of the old Army Corps employees were being replaced with bright young men (mostly) at the beginning of their careers. Peter Hubbell hired Mark, a professional engineer. It is rare to hear someone mention Mark without Pete or Pete without Mark. They were so perfectly matched that together that they made a powerful team. Mark was aggressive and gregarious, whereas Pete was more laid back and thoughtful. Mark was a hard charger on the outside, protecting a much softer center. Pete had a softer outside with a steely center. Pete had excellent political instincts, while Mark was a strategist. From the moment they met, *they just clicked,* in Mark's words. It was clear that Pete was being fast-tracked to the top of the District.

In only a few years, Hubbell was two levels above his old boss in the District hierarchy. Mark followed him. By the late 1980s, Pete was the assistant executive director with a clear ticket to the top. Gary Kuhl, the executive director, liked Pete and wanted to leave the agency in good hands.

"In his last three months, Gary let me run the District," Pete said. "It was great on-the-job training. I knew what I was getting into. By the time Gary left, I knew what was going on, and I knew I could do it."

The governor of Florida appoints the Governing Board members of each District. The Florida Senate must approve them, but in those days, few appointments were challenged. Board members served a four-year term. The board hires the District's executive director. Each time the governor changes, the Governing Board can change. Some transitions were more traumatic than others.

Former District chairman Derrill McAteer said that he was inclined to relate policy preferences to the personalities in place, not the political party. Pinellas County was among Florida's wealthiest urban communities, and it was at the forefront of the political contest over water-supply development. In 1980, with McAteer gone, the new, urban-dominated Governing Board didn't want to fight with Pinellas and St. Petersburg and sought a more collegial relationship. At the same time, both the Florida Legislature and the Department of Environmental Regulation were delegating more responsibility to the District. With its independent taxing authority, the District had the financial resources to do a job without further draining the state's general fund. At the time, the ad valorem taxing authority of the District had two parts. Half of the allotted one mill (fifty cents of each tax dollar) can be levied by the Governing Board to pay for operations and other directed activities. The Southwest District had basin boards that oversaw projects in a more localized area. The basin boards had no regulatory responsibility. The District's local basin boards could collect the other half mill for local projects.

From this perspective, it is possible to see the cyclical nature of the relationship between the District and the Authority. When the relationships among leaders were good, the staff was expected to cooperate. When the relationship was less cordial, the technical staff tried to stay out of it and focus on the immediate task at hand: data collection, interpretation, and application. Some people suggested that District executive director Gary Kuhl was especially sensitive to the fact that the consequence of local governments' wrath might be his termination. They could demand his resignation or dismissal and had sufficient political power to insist on it to the governor.

On the other hand, some say that cooperation with local governments is the right thing to do. Like so many facets of public policy, beliefs are a matter of perspective and values. The public interest is established based on how you draw the circle and who is inside it. The fact remains that Gary Kuhl successfully maintained a cooperative relationship with the Authority.

It was in one of those periods of cooperation between the District and the Authority that Freeman Polk challenged a District decision. Polk was one of the few water activists who had both commitment and money; Polk sued the District over its plans to pump the Cross Bar wellfield in central Pasco. He was concerned that pumping there would affect the value of his nearby property, which was wedged between the Cross Bar and Cypress Creek wellfields. He was concerned about his cattle ponds drying up from pumping. The District believed that Cross Bar was a good, upland site for pumping and was supported by the Authority and its outside counsel, Ed de la Parte.

In short, it was the District and the Authority against the activist farmer. The District took the position that the cattle ponds did not constitute an "existing legal user," a factor that could have affected the Pinellas permit at Cross Bar. Florida's laws say that a new user cannot impact an existing legal user. Because of the District's position, Polk had no legal protection.

In 1985, the District hired a new hydrogeologist named John Parker. John had finished the coursework for his master's in geology. By 1988, while the Freeman Polk case was moving toward court, Parker was still too new to be an expert witness but was senior enough to attend as an observer.

"I was concerned about some of the points the Authority was making at the time," Parker remembered. "I talked to the attorney representing the District and told him that these would be a problem for us later."

The District was doing a multi-wellfield study that was already demonstrating some connection between pumping and wellfield impacts. John was concerned that a total win in the case would make it hard to manage future cases—cases where the District might oppose a local government.

"I was told, 'Our objective is to win, not to win with limitations,'" he said. In any case, the studies demonstrating possible effects were stopped. "It just petered out, never got finished," Parker said. No one knows why. Freeman Polk lost the case when the hearing officer, Don Conn, ruled against him. By 1995, Don Conn was general counsel for the Authority.

John Parker grew up in an environmentally aware household. His father was a forester, a real "green guy," back when the environmental movement was first gaining momentum. John admired his dad for his work protecting trees and forests. He was surprised and hurt as a kid when others told him his dad wasn't "green" because he worked for a big paper company. John's father grew trees to cut them down, and to some, that made him a bad guy.

John knew better, and his rigid sense of what was right and wrong was connected to his defense of his dad and the conviction that it is essential to defend what is true. His values kept him loyal to his science, to the District, and the cause of public resource protection—regardless of the politics. Over the years, he found confederates who shared his long-term view.

"We just lock arms, continue to generate data, and get ready for the next time," he said, smiling. In the meantime, John carried no debt, so if he was fired for his work or his positions, "I guess I'd become a rich man," he said, smiling, the implication being he would sue for wrongful termination. By keeping no debt, he ensured that his steadfast fidelity wouldn't harm his family or his personal beliefs. John found ways to balance the responsibilities of his position, the direction of his superiors, and the demands of his conscience. It was a delicate balance. The Freeman Polk case was John's first exposure to hydropolitics.

Although the multi-wellfield study was never completed, the District continued to inventory water use and resources and to issue water use permits. The District staff was beginning to piece together the specific relationship between taking water from underground and lowering lakes and wetlands on the surface in this region, irrespective of the drought.

The staff believed that the Floridan Aquifer had been overdrawn. The Floridan Aquifer is a spongy layer of limestone deep below the ground where freshwater is stored naturally. It provides 80 percent of the public drinking water for West Central Florida. The aquifer provides drinking water, and it is replenished by rainfall. Depending on the hydrogeology of an area, the aquifer sometimes breaches the surface (in rivers, springs, or other water bodies), and sometimes it doesn't. The greater the connection between the surface and the aquifer, the greater the impact of drawdown.

"In the 1980s, it was the District that put the Starkey and Cypress Creek wellfields in place," said Hubbell. They bought a lot of land, then encouraged the Authority and member governments to put the wellfields there. From further down in the organization, it looked like Gary Kuhl did whatever local governments wanted, gave them whatever they asked for, and provided them with whatever political cover was necessary. It didn't sit well with some District staff.

Even so, what might have seemed cooperative then—the District providing land for new water supplies—is part of the history of the two organizations and the member governments. It's a history with real consequences for people and the environment. It is a history that colors the organizations' relationship with each other and the people who worked for years on water-supply problems in Tampa Bay.

Gene Heath is one of those people. A technician by training, Gene has always been interested in environmental and engineering matters. When he was selected to be the general manager of the Authority in 1978, he looked forward to the difficult challenge. Gene heard all about tough Pinellas commissioner Rainey and the contentious fights over water. He thought that by being honest and focusing on new water supplies, he could avoid some of the political pitfalls.

"I always tried to work with District staff," Heath said, admitting that there is a natural tension between permitter and permittee. With one eye on the environment and another on water-supply development, Heath tried to walk a fine line. He tried to bring new water supplies online. Cypress Creek was one.

The Cypress Creek wellfield was supposed to be a sign of cooperation between the Authority and the District. In Heath's words, the District said, "We'll buy the land, and you put the wellfield here." The District was committed to flooding the property, anyway, virtually guaranteeing that the wetlands would stay wet and the water supply would be renewed since those wetlands would recharge the aquifer—the drinking-water source. Still, getting the permit was no easy matter.

Ed de la Parte remembers a particularly tough meeting with Derrill McAteer (self-described as hardheaded), clearly unhappy with the proposed Cypress Creek permit.

"He would hand the gavel over to someone else, then step into the backroom for a private conversation," de la Parte remembered. "Fran Davin was a Hillsborough County commissioner and chair of the Authority. At one point, McAteer came back to his seat and said, 'We're not going to pass this thing; we're going to defeat this permit unless you do this, and this, and this.'"

In a brief exchange, Davin asked de la Parte if legal action was pending on the case. "It is," he told her. "Fine," she said. "Would you please inform District counsel that we'll do our talking at the podium?"

"McAteer turned red and started yelling," de la Parte said, a broad smile spreading across his face. "We left." Litigation was pending.

McAteer didn't remember the meeting at all but did remember being very concerned about the Cypress Creek permit. "I can tell you," he said, "Fran Davin and I always had good communication. It never broke down, not once."

On the way out of the District offices, Davin and de la Parte found Gene Heath, the Authority's general manager, with his head between his knees, hyperventilating, worried about whether the permit would go through. The permit did go through, passing by one vote cast by Bob Martinez, the mayor of Tampa who would later be governor of Florida. Heath said, "I was there for the vote, then left immediately after. It was a tough meeting. Why stay? If you say thank you to the board members who voted for it, you rub salt in the wound. It's better to take yes for an answer and go."

Behind the Curtain

The Cypress Creek wellfield began production as planned, but the flooding never happened. Some said that scientists in the District believed that "flooding would be worse than pumping" for the environment. Authority ecologists objected as well. "It was a beautiful system out there," Heath recalled. "No matter who went out there, they came back and said, 'Don't flood it, it's too beautiful to lose.'" So, the District decided not to flood it. That choice was part of the District's decision to stop construction of the Four Rivers Basin project—the project that was meant to manage flooding and for which the District was established. Policy, guided by science, was moving toward nonstructural flood control. Structural flood controls are big construction projects like the Tampa Bypass Canal, wherein constructed channels and controls determine the rate and path of water flows. Nonstructural controls include preserving large tracts of lowland so that excessive rainfall can be captured and can naturally recharge the aquifer. As Florida grew, it became more difficult and costly to secure the necessary land. Cypress Creek was a planned regional drainage area for the Four Rivers Basin project. The floodwater detention structure is still out there today and has never been used.

One morning, water suppliers and managers awoke to find Ted Rochow, a District wetland ecologist, quoted in a *Tampa Tribune* news story. There was a picture of him gesturing to a place on a tree in a wetland near the Cypress Creek wellfield. He told the reporter that the mark on the cypress tree was not natural. Water levels were far below what they should be, and soil had begun to subside, exposing the roots. It was not due to drought. He said the wellfield caused the problem.

Representing the Authority, de la Parte remembers the incident. "We couldn't believe that Ted was in the paper about the wellfield," he said. "The District insisted that we put the wellfield in, and then this?"

The Authority wrote a letter to the District demanding the story be corrected. "The letter said something like 'This is crazy. You were going to flood Cypress Creek, and then you didn't. You said you wanted to monitor the effects of a wellfield on cypress swamps, and so our requests [for augmentation] were denied,'" according to de la Parte. "Gary Kuhl directed Pete Hubbell to get together with the Authority to work out an augmentation and mitigation schedule."

Then Kuhl wrote a letter to the Authority to smooth things over, saying, "You are right, the Authority's not at fault. We need to look at options and facts."

Ted Rochow worked for Mark Farrell. Farrell was called into Gary Kuhl's office. Hubbell was already there but didn't say much. According to Farrell, Gary wanted Ted Rochow disciplined. He wanted Ted written up, a letter put in his permanent file.

"For what?" Mark asked. "Was he wrong? Was he lying?"

"That isn't the point," said Gary.

"Help me out here, then. What'd he do wrong?" Mark asked.

"I've gotten calls," Gary admitted.

Mark assumed that meant Governing Board members or Chuck Rainey, the powerful Pinellas commissioner known to demand the ouster of District or Authority staff not to his liking.

"I looked at Pete," Mark said, "and he kind of threw his hands in the air as if to say, 'What can I do? He's the executive director.'" So Mark went to his office, wrote the letter of reprimand, and took it to Pete, who approved and forwarded it to Gary Kuhl. Kuhl signed off on the letter, and it went back down through the chain of command to Mark.

He threw it away.

But a letter did appear in Ted's personnel file, dated March 27, 1986, and labeled "confidential." The letter says that "on March 13, 1986, Mr. Rochow was interviewed by a member of the news media in regards to a current District permitting activity. In response to that interview, Mr. Mark Farrell, Director, Resource Management Department, informed Mr. Rochow of the District procedure regarding media interviews. On March 20, 1986, Mr. Rochow was again interviewed by a member of the media regarding a District activity and followed all District procedures regarding such interviews."

The memo was signed by Pete Hubbell.

Back then, facts were in painfully short supply. While McAteer had persuaded the Governing Board to invest in technology in the 1970s, upgrading was slow and limited. It was in the late 1980s when board member Roy Harrell, a St. Petersburg lawyer, demanded to understand the extent of the environmental problem. Roy had already pushed the staff to increase conservation efforts and to encourage the local governments to do it, too.

According to Farrell, Roy said, "Listen, I don't mind spending a gazillion dollars on conservation, but how big a problem do we have? Will we have to spend $50 gazillion or $1.98? Define the problem for me." The staff's kneejerk reaction was "Well, it's gonna cost a bunch of money." They needed new technology, they needed outside help, and they required board support.

Democratic governor Bob Graham initially appointed Roy to the Governing Board in 1985, although Republicans and Democrats alike eventually reappointed him. Roy was a big man with a big heart. Some called him a tax-and-spend Democrat. Roy's philosophy was always "If you're going to make a difference, make a big difference." His father had told him that when you leave this world, it is only the footprints in the sand that remind people of your presence, your contribution.

Roy was a real estate lawyer working in his firm in a tall building in downtown St. Petersburg. Unpretentious, Harrell was known to bear-hug friend and foe alike. This can be either a physical or a political matter because he wouldn't take no for an answer. Some called him pigheaded, but Roy never lacked courage. He served thirteen tough years as a member of the Governing Board, six as vice-chair, two as chair. During that time, there were many difficult decisions, and Roy never flinched from one.

"There was a lot at stake," Hubbell said. "Without the data, we could not come up with management strategies, regulatory controls, incentives, or anything. We began to realize exactly how adversarial the Authority and its members would be about change." It was the same problem faced by McAteer in the 1970s when the Florida legislature first gave the District regulatory authority. However the board didn't register the pending problem with local governments. They just knew that the staff said they couldn't define the problem without proper funding.

Farrell added, "Roy asked, 'How much? Go find out and tell us.'" So we did, and we told the board it's going to take five years, a lot of staff time, a lot of consultant time, and millions of dollars. But they approved it. So the District started the WRAPs [Water Resource Assessment Projects], in which they defined an area, analyzed the hydrology, the natural systems, the (potential) sources of water, the impacts, and everything else."

This began the first broad-based, scientific, systematic study of the water resources of the region.

"We needed to know how much water there was and how much water was available," Hubbell said.

"Without that," Farrell added, "when the time came in the early 1990s, we'd have never been able to do what we did." Others argue that when the time came, the District still wasn't ready to do what it did—Roy Harrell among them.

Chapter 4
Conflict Spreads

The Southern Water Use Caution Area

The problems faced by the District in the late 1980s and early 1990s were not limited to Tampa Bay, where the staff believed that environmental impacts were related to water-wellfield pumping. Another region of the District (5,500 square miles) was dubbed the Southern Water Use Caution Area (SWUCA), which had a different set of problems. Unlike the highly urbanized Tampa Bay region, the SWUCA had thousands of small and large agricultural users, miners, and public water suppliers. The SWUCA

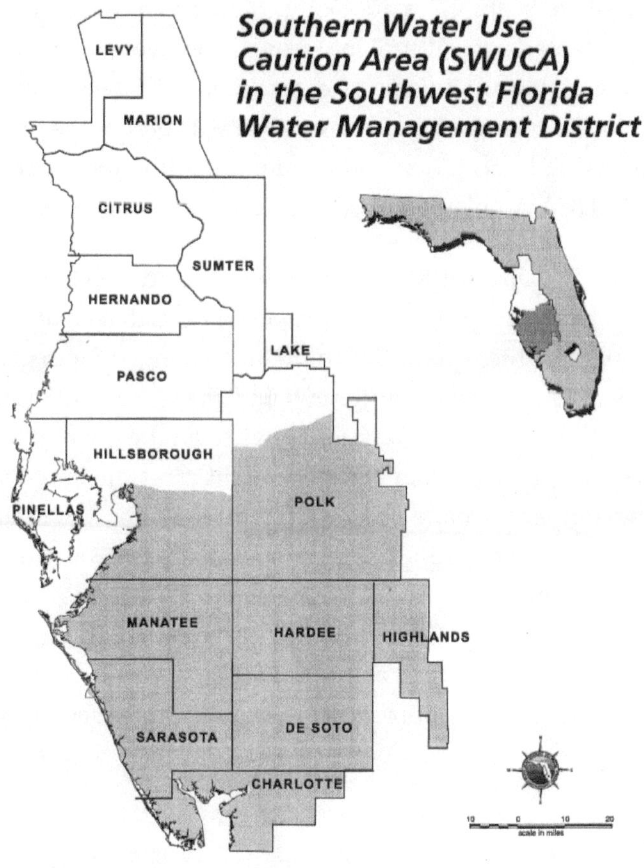

had a different hydrogeology, one that was more regional than northern Tampa Bay, so a local management plan wouldn't work.

The problem demanded a plan that encompassed eight counties. To begin a stakeholder engagement process, the District held a series of public meetings where technicians presented information and solicited input from locals. The meetings were brutal, according to some observers. District executives took public criticism for months on end. In those meetings, they admitted that more than one billion gallons of water had been permitted, that only half of that was being used, and that even that amount was causing a problem. They showed people the "red-hole map," which they claimed identified the area of most significant impact. It was in the region of southern Hillsborough and northern Manatee Counties known as the Eastern Tampa Bay Water Use Caution Area.

"They just took a beating," said Doug Manson, a water lawyer who primarily represented agricultural interests. "They were taking a lot of hits as an agency and as individuals," he recalled. "They were trying to do something new, and people hate change. That was part of the problem. But they kept gathering information and coming up with more alternatives; as a result the SWUCA problem kept changing."

Assistant Executive Director Mark Farrell was the architect of the solution to water problems in the SWUCA. He proposed rules that created a de facto market for water in which quantities could be exchanged for cash or credits. At the time, it was remarkable and new, and many people felt the courts wouldn't uphold it. In Florida, water belongs to all the people, so allowing a private interest to benefit from a permit might cross the line. Then again, private water companies make a profit from public resources in the form of bottled water and even public supply, but a separate bureaucracy (the Public Service Commission) regulates the economic aspects. The SWUCA alternative proposed by the District had a regulatory component that some interests thought wasn't strong enough. Farmers didn't trust the District and weren't willing to support rules that gave bureaucrats even more authority.

Manson pointed out that the District had many meetings and each time blamed a different cause. "One time it was saltwater intrusion, the next it was upward migration, then it was a shortage of water," Doug said. "It was funny and sad."

Doug Manson is better liked than most lawyers, although he always opposed the District unless his client's interests happened to align with District objectives. It didn't happen often. Oddly, though, District staff held him in high regard. Focused mostly on agricultural water interests, Doug would rather find a way to work out a problem than expand it. He was a nice guy who represented his clients well. He did not employ the take-no-prisoners attitude of many lawyers and engineers who challenged the District. He distanced himself from Ed de la Parte's scorched-earth approach. Doug managed to build a successful career and a reputation as a nice guy at the same time.

By the early 1990s, Sonny Vergara was the general manager of the Peace River Manasota Water Supply Authority. Like the West Coast Regional Water Supply Authority, but younger and less formally organized, the Peace River Authority included Sarasota, Manatee, Charlotte, and Desoto Counties. Vergara had left government service for a time after his experience at the St. Johns River Water Management District, and he consulted with the Peace River Authority while trying his hand at growing and importing flowers.

He returned to government service at the Peace River Authority to help ensure cooperative water-supply development among local governments. Competition for water south of Tampa Bay was different. Sarasota was the fastest-growing community, with beautiful and abundant beaches and no water supply to speak of. Sarasota had gone to brackish-water desalination years earlier. It was an essential, if expensive, move. Like Pinellas County, Sarasota County rapidly outgrew its sources and began to look elsewhere. Sarasota looked north to Manatee County, which had the Manatee River, and south to the Peace River in Charlotte County. Wealthy Sarasota, like Pinellas, had the financial resources, political muscle, and commitment to get the water it needed.

There had been some rough times among local governments in the south, but Charlie Black, the Governing Board chair (1992–94), delivered a large District check—$1 million—that was enough to persuade the local governments to work together through the Authority. It was enough to stave off conflict for a while.

In the meantime, Vergara attended some of the early SWUCA meetings and didn't like what he saw. In public meetings, farmers and others would scream at Hubbell, Farrell, and Dave Moore, assistant executive director for resource management. They were accused of being incompetent, dishonest, and a host of other things. Sonny watched these meetings and related to the criticism based on his experience in the St. Johns District.

"I saw the baggage building," he said; his departure from the St. Johns District still stung. "They did a great job of controlling themselves, but it was a terrible thing to watch."

More People, More Water

By the late 1980s, the growth in Tampa Bay was evident everywhere. Roads were choked with cars, farmland was becoming rooftops, and the demand for water was growing. Between 1980 and 1990, the West Coast Regional Water Supply Authority brought some new water sources online, but they were insufficient to keep pace with growth. The permits issued previously were inched upward and renewed. The Authority board refused to approve any new water source except groundwater. Pinellas County blocked all other potential alternatives. Pinellas commissioner Rainey insisted that there was plenty of groundwater and that it was cheap and accessible. New supplies couldn't be developed without Pinellas funding.

As complaints from citizens living near the wellfields began to increase in response to environmental impacts and collapsed private wells, Hillsborough County and especially Pasco County started to object publicly to pumping. Both the District and the Authority insisted that problems surrounding the wellfields were due to lower rainfall levels—not to wellfield pumping. The continued conflict among the Authority member governments and the veto power of each board member made developing new supplies virtually impossible. The persistence of individual self-interest and the structure of the Authority governance stymied progress. Despite the theoretical advantages of collaboration, water-supply development was stalled. In the meantime, the combined population of Pasco, Hillsborough, and Pinellas Counties increased by 30 percent between 1980 and 1990. And there was the persistent drought. The Cypress Creek wellfield was producing water, and production was increased at any facilities with additional capacity, including the Cross Bar wellfield in central Pasco. Permits are issued with conditional levels, meaning that pumping can start at one level and, if there are no problems, be increased to the next. In this case, it became evident that while one level might have acceptable environmental impacts, the next level would not. That's why permits have conditions that allow them to be revoked, recalled, and modified.

But it takes political will to change things.

The Governor's Board

In 1988, Republican governor Bob Martinez was in the Governor's mansion. He appointed a new member to the District Governing Board who took a seat traditionally held by Pasco County—the primary water-donor county—but the new appointee resided elsewhere.

The Pasco County Commission was outraged. As the major water donor, it felt entitled to a permanent seat. During the legislative session, there were bills drafted to change the board's composition. Every community wanted to make sure it had a permanent voice on the panel. On June 30, 1990, the sitting members were officially off. Martinez reappointed them and, through the new bill, added two additional seats.

"The problem was they had nine good old boys, and they added two more," Hubbell recalled with no particular rancor. In November 1990, Martinez lost the election to Democrat Lawton Chiles, and the board might have changed again. The District staff were reeling from the sudden, multiple changes. They didn't know what to expect. To ensure a small bit of continuity, they lobbied the governor's office to leave Pinellas County resident and lawyer Roy Harrell on the board.

"We tried to keep Roy," Hubbell said. "Roy was a strong board member . . . we tried to keep as many as we could, but the politics were very tough." The new governor came from an agricultural background, and farmers were a critical constituency for him and the District. Farrell and Hubbell and their new plan in the SWUCA might be in trouble. Eventually, Harrell was allowed to stay. He was a Democrat. Sally Thompson was also allowed to stay. She was a woman, an environmentalist, a Tampa employee, and a Democrat. Hubbell admits that despite his best efforts to influence the governor's appointees, he was mostly unsuccessful. Roy, Sally, and Charlie Black stayed. The rest were replaced. This left very little institutional memory on the board. The staff was forced to start the education process all over again.

"It was around this time that the *New York Times* ran a big story about waste and abuse at the water-management districts," Hubbell recalled, shaking his head. "It was mostly about the *South* Florida Water Management District, but that didn't matter; we all got painted with the same brush."

Later, when the District was in conflict with public suppliers and trying to change policy, all of the districts found themselves facing increased scrutiny and, yet again, "painted with the same brush," though this time it was about resource management.

At one of the early meetings of the new Governing Board, Chiles appointee John Hamner decided to make a point. A retired editor of the *Sarasota Herald Tribune*, a *New York Times* paper at the time, he walked in with about twenty copies of the *Times* draped over his arm like the napkins of a headwaiter.

"He put one at every board member's seat," said Hubbell, smiling at the memory. "I thought to myself, 'This guy is going to be trouble.' I called that one right, didn't I?"

Joe Davis was another new board member. A Chiles appointee who had a longtime relationship with Lieutenant Governor Buddy McKay, Davis had both an agricultural and a real estate business. He was also a third-generation native Floridian. In 1990, when he joined the board, there was a recession. Trained as both a lawyer and an accountant, Davis thought the annual 6 percent increases in District staff pay were outrageous. He believed a tax cut was a good faith effort for the District to demonstrate fiscal responsibility, and he was committed to making some changes.

"Since the governor had changed eight of eleven board members, we thought that was a clear signal he wanted some changes. We also thought that meant more consideration for the environment," Davis, a Democrat, said.

"Joe Davis came in and was very aggressive," Mark Farrell admitted. "He questioned everything we did. It was as if Governor Chiles said, 'I don't know what's wrong, but you go in and fix it,' and Joe was leading the charge. We were shaken, scrambling to prove ourselves."

Roy Harrell defended the staff.

"Roy spent a lot of personal capital defending us," said Hubbell. "He'd been there long enough that he had a historical perspective and his perspective. He defended the positions we'd taken, he defended where we were heading, and the rest of the board beat on him unmercifully. He never won, but he never failed to jump into the fray."

"You couldn't ask for a better ally," said Farrell. "He would pull every chit. There were board members who only did what was good for them. But not Roy. He stood up." Harrell said that defending the staff was more than just caring about the people. "It was one of those 'new board' things," Hubbell agreed.

"It's human nature, you know," Hubbell said. "To come in and think you can fix everything, that what's going on is wrong, and your ideas are right. Then you look at the District and the situation with the [West Coast] Authority and . . . there's a whole story. But you don't know it. Roy tried to help there, but it still took two years to gain [the board's] trust." Earning their trust meant figuring out how to do things they wanted without incurring too much damage to the long-term plan.

"We were sitting down at Fat Willy's near the Tampa Service Office," Hubbell said. "It was Joe and me and Mark. Joe said, 'Look, we're gonna find out what sustainability is. In the meantime, we need to raise taxes to get the money to pay for new water supplies.'"

"Joe was so pleased with the idea," Hubbell said, but it brought a cold sweat to the two executives. "Mark and I were like deer in the headlights. When Joe left, we were like, 'He's crazy.'" Still, they had an obligation to explore the idea as an option—and they did.

Sustainability describes the use of a resource to such a level that it can replenish or renew itself on an ongoing basis. In water, it means ensuring that no more water is taken than can be restored so that the primary source is not mined, not diminished. At its heart is the notion that our resources are not limitless. As tough a concept as it was to sell on the outside, inside the District, it was difficult as well. Unlike the other water-management districts, Southwest was broken into smaller basin units. The ad valorem taxing authority was divided equally between the basin and the Governing Board.

In the mid-1980s, the Governing Board chair moved to do away with the basins to get control of the entire millage allocation. The District boards went to court with each other. The basins remained, as did their taxing authority, but the relationship between the Governing Board and basin boards remained strained. Each Governing Board member served as ex officio chair or cochair of the local basin board. The Governing Board members voted on local projects only in case of a tie. The basin boards were needed to support a tax increase.

"Roy supported Joe," Farrell pointed out. "Roy had been raising taxes at his basin board to fund projects, so when Joe proposed the New Water Sources Initiative, Roy said, 'Yeah, challenge grant,' and off it went."

"Roy told me if you're going to do it, make it big enough to make a difference," Davis said. "It was my idea, but Roy was 100 percent there, and the rest of the board was, too." No one realized the heat the District would take over the tax increase.

"But Joe, God bless him," Farrell remembered, "would take the floor at the public meetings and go through the reasons for raising taxes. He didn't try to lay it off on staff; he took the questions himself."

"We've been lucky to have people who kept it all together and stood by their sometimes-unpopular convictions," Hubbell noted. "The Roy Harrells, the Joe Davises... There've been others." Almost everyone involved at the time remembers the charts Davis used. The jagged red line was heading up, then down, and up again, with the new money earmarked for new water-supply projects.

"People showed up with hatchets, you know to chop taxes," Hubbell said, a big smile at the memory. This kind of meeting is much better as a memory.

Existing Tampa Bay Water Supply Facilities

1998 Water Year
Average Production - MGD
Major Water Supply Facilities

Tampa Bay Water	
Central System Wellfields	
Cypress Creek Wellfield	23.2
Cross Bar Ranch Wellfield	20.8
Cypress Bridge Wellfield	7.6
Northwest Hillsborough Wellfield	8.5
Cosme-Odessa Wellfield	11.0
Section 21 Wellfield	10.4
South Pasco Wellfield	14.2
Morris Bridge Wellfield	13.0
Eldridge-Wilde Wellfield	22.9
Other Wellfields	
South-Central Hillsborough Wellfield	19.5
Starkey Wellfield	11.9
North Pasco Wellfield	1.4
Surface Water Supply	
Tampa Bypass Canal	5.5*
City of Tampa	
Hillsborough River System	67.3*
Total	231.7

* Hillsborough River total includes Tampa Bypass Canal quantities.

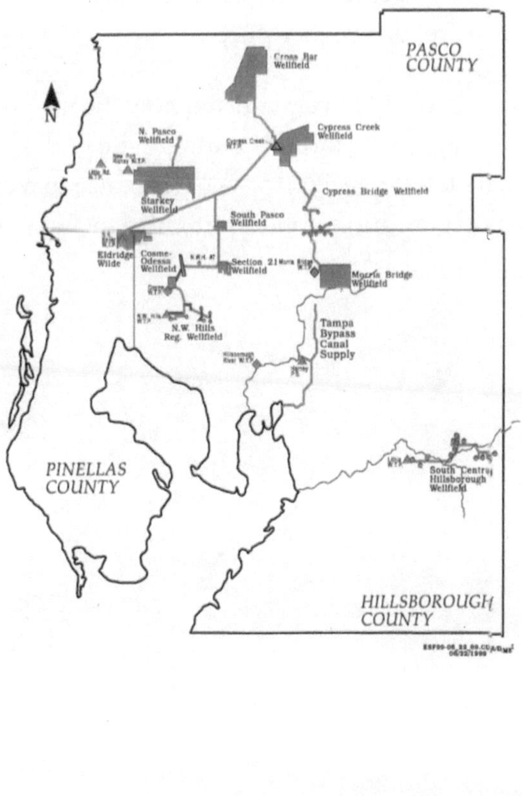

"I was naive, I guess," Davis said. "I thought if we cut taxes to get rid of the waste, we could raise them later," but it didn't work out that way. That meant that the District got virtually no credit for cutting taxes but took a lot of heat for raising them.

"There seemed to be no light at the end of the tunnel," Davis said. "No one was creating any new water, so if the District could step in with some money to encourage water-supply development, it seemed like an alternative to suing each other. We were sure user fees for water would be unacceptable. We needed the money, and even if it might not solve all of our problems, it might take the edge off."

Each of the people in this situation was trying to serve a vital public interest: provide a safe and affordable water supply while protecting the environment. Both staff and board members stood and faced angry taxpayers, businesses, political interests, and others because they believed that in the long run it would pay off for the entire region. Board members were thinking about the greater good. Staff was thinking about how to get it done, scientifically, pragmatically, and politically. They withstood the pressure for a higher purpose.

No one could have predicted how the new money generated by Joe Davis's idea and Roy Harrell's enthusiastic endorsement would make it possible years later to facilitate the best chance to end the water wars in the Tampa Bay region and provide water to a growing region.

Chapter 5
One to the First Power

People come to the Tampa Bay region from everywhere. They choose the life they want: the vibrant St. Petersburg waterfront with its restaurants and bars, Tampa with its access to sports and entertainment, beachfront Pinellas, or rural Pasco and Hillsborough Counties, where you can see neighbors but not hear them. Mostly, every location comes with great weather and the promise of a better tomorrow.

One of the newcomers to rural central Pasco County was Steve Monsees. A retired Green Beret colonel, Steve had completed a tour of duty at Tampa's MacDill Air Force Base and liked the area. To hear him tell it, Pasco County had the quieter way of life he longed for—it had that hometown feeling he wanted his son to know. But what convinced him to stay was the water.

In 1988, Steve bought a seven-acre parcel in central Pasco that would become a retirement paradise for him and his family. It was a beautiful location with a hundred-acre lake out front and a pond in the back. He dreamed of great fishing and a settled, peaceful life for his family. There was also room for horses. Steve and his wife looked forward to enjoying their retirement. Though he was on assignment in Africa, he would periodically visit MacDill and his property. He noticed changes, like the lake levels dropping, but didn't think anything of them since periodic drought happens everywhere. In 1989, Steve began construction on his dream home.

By the time they moved in, Christmas 1991, the hundred-acre lake in the front yard was nearly gone. He was concerned because he'd heard his neighbors talk about the "effects of Cross Bar wellfield," but he didn't pay much attention. When water levels continued to drop, he suspected that the problem was something more than a drought.

As a career Army officer and a Green Beret, Steve's experience with government was different from that of his neighbors. The Water

Management District was not the enemy; it was an ally—there to serve the public. He turned to it for answers.

But he began with the Authority. A woman whom he could barely recall had answered his questions. He thought her name was Penny. "I even thought she cared," Monsees said. But Penny claimed that drought caused the problem. When he asked more questions, he couldn't get a straight answer. "They blamed it on development; they blamed it on rainfall. They blamed it on irrigation. It was everything but the wellfield," he said, disgusted. "It was obnoxious, and I got mad." Steve remembers that his first calls to the District and Authority were made in 1991 or 1992, but his first documented correspondence is dated September 10, 1993. It read in part:

The purpose of this letter is to complain about the excess water production at the Cross Bar Ranch Wellfield, which has resulted in the total loss of water in my pond, destruction of wetlands on my property and increased the likelihood of sinkholes. Until 1999 my small lake flourished with fish and other wildlife and along with the other lakes and ponds was my single most important consideration which I purchased my land here. By the summer of 1990, my pond was dry . . . I was informed . . . that water loss was because of a long period of low rainfall. We have since learned by review of public documents . . . that the principal reason for the loss of water from my pond and the other ponds and lakes . . . results from excess water production at the Cross Bar Ranch Wellfield.

Steve and his neighbors were disgusted with the Authority, the District staff became their friends. The staff were a great source of technical information. "They were professionals," he said. "They answered our questions, and they weren't one-sided."

Steve remembered the help of District staff but one employee stood out: John Parker. John Parker worked in the regulatory section of the District's Brooksville office. He was a geologist with a calm attitude and impish smile most frequently displayed when he was about to "lob one in" on some environmental bad actors.

"The first time I met him, a bunch of the neighbors went with me. We just called up and made an appointment."

Steve returned to the District on many occasions to get briefings and information from hydrologists, geologists, environmental experts, biologists, and engineers. "Professional," he said; "every one of them."

Slowly, Steve began to understand the available technical information. He came to know the hydrogeology of the area, the engineering, the installation of the wellfield, and the effects of drought. The more he learned, the angrier he got. Steve became convinced that wellfield pumping had sucked the water out of the ground, taking with it surface-water bodies, including the hundred-acre lake in front of his home and the pond behind it. Left behind was dry and dusty open space where nothing lives, the view best handled with closed curtains.

Steve blamed the Governing Board for refusing to see the problem. He and his neighbors believed they were a remote sacrifice for powerful political forces in Pinellas County, where the water was going, where the grass was green. "Our property is ruined," he said, "while lawns in Pinellas are lush and green." He couldn't see the sense or the fairness in that. "I just wanted them to do their job," he said of the District Governing Board.

Steve wasn't alone in his feelings—technical professionals inside the District shared his frustration. They thought that the Cross Bar wellfield had been a mistake, that it had been overpermitted—given permission to take too much water—that the impacts from pumping had been grossly underestimated, and that something should be done. The first reliable data linking wellfield pumping to environmental impacts were coming in strong. Technical staff felt they had what they needed to prove in court that harm had taken place.

Harm was the official standard that allowed them to take regulatory action to change things. But in 1993, the staff weren't sure the Governing Board was ready to hear the harsh reality. In fairness, the staff had put together a six-month program to educate the new board members on their scientific findings, and that process had only just begun. But to Steve and his neighbors, whose lakes were dry and homes devalued, the slow and steady pace was simply the politics of delay. In the fall of 1993, Steve was ready to do something. He decided to talk to the Governing Board himself. He told John Parker about his plan.

With John's help, Steve Monsees changed the face of water-supply policy in Tampa Bay. "I was told that on this day at this time, this subject matter will be discussed," Steve recalled. "I was told that if I addressed the board, then they would hear me."

The Wave of the Future

It was a cool, dry morning on January 26, 1994. I'd been at the District for less than sixty days. Hired when public polling showed support at 23 percent, the job was to be, as a friend put it, "a make-or-break-your-career" opportunity.

At the time, I thought my biggest problem was the $300,000 Do Your Part water-conservation campaign that was behind schedule. Or that an investigative journalist from Channel 13, a local CBS affiliate, had been on site since October interviewing staff and researching records, and no one had told me. Since February ratings help determine advertising revenue, stations often hype their stories to entice the maximum number of viewers. That investigator, his activities, and my District-staff management challenges were high priorities. But not for long. While I was at my desk working, John Parker and Steve Monsees were meeting in an upstairs conference room.

District executive director Peter Hubbell and assistant executive director Mark Farrell were in the Governing Board meeting as was Bill Johnson, the St. Petersburg utility director. Water lawyer Doug Manson was in the lunchroom, watching on the closed-circuit TV. Doug was there to represent an agricultural client on a water-permit matter. He was waiting for his turn to address the board. Some people remember Steve Monsees's Governing Board presentation as the last item of the day. Some remember it occurring during or just after lunch. Some remember a full boardroom, while others can't remember why they were practically alone in the room. Regardless of where they were when they heard it, Steve's presentation struck a chord with everyone.

Utility director Bill Johnson is a nice man with a gentle nature and an easygoing personality. He fit in well with St. Petersburg's city management, who tended to think of themselves as long-term planners, as leaders, as good guys. Bill had great personal regard for the environment. As a devout Christian, he said that the Lord demands he be a good environmental steward. Bill was proud of the conservation efforts of St. Petersburg, his high standards, and the quality of water he delivered to city customers. His water won taste awards just about every year, he said. The St. Petersburg water-use permits had been under review and subject to reissuance since 1992, nearly two years earlier. As part of their condition for renewal, permits must be reviewed every few years to ensure that no unexpected problems have resulted.

By January 1994, those permits still hadn't been reissued. The length of time was very unusual for public-supply permits, which are generally quickly reviewed and reissued. St. Petersburg's permits had been in place for decades. The District had some concerns about the permits, and they were taking time to negotiate mutually satisfactory solutions. The District had not reissued them yet, but Bill was confident that they would be in due course. They always had been; why not now?

"Staff made presentations to the board," John Parker said. "The board shared our concerns about the environment, but nothing was happening. We needed Steve to get right in front of the board with 'in your face' information."

The meeting between John Parker and Steve Monsees didn't take long. Steve had written his comments but didn't share them with John. What Steve wanted to do was show John the pictures he intended to show the board. Steve wanted John to help him figure out the priorities for his presentation. In Steve's mind, this was the most important talk of his life, so he wanted to get it right. He wanted to use his three minutes well. Three minutes. That's how much time the public are given to plead their case before a board that has the power to change their lives, their livelihoods, and their future. According to John, he offered a little advice. "'Use this picture, maybe that one.' 'Leave it up; fewer is better.' 'Keep it simple.' That kind of thing. He didn't really need my help," said John. "He had a powerful story to tell."

Steve Monsees was a big man. More than six feet tall, he was well designed for the missions he served in Vietnam, Africa, and other places that Green Berets go but never talk about. At the time, Steve was tipping the scales at an intimidating 240 lb. Still, his commanding presence and giant size belied his gentle nature. When he talked about his role in the water wars, there was seething just below his civil surface.

For more than an hour beforehand, the Governing Board bickered about the augmentation of St. Petersburg's Section 21 wellfield in northwestern Hillsborough County. Augmentation is the practice of pumping groundwater or reclaimed water into a surface-water body or wetland to keep water levels up. At best, it's a cosmetic cover for an underlying issue. One board member, Margaret Sistrunk, was angry. "When," she pleaded, "are we going to have some real answers? When are we going to quit issuing permits? When are we going to do something about what's going on near these wellfields? And how can we be so cynical as to keep telling people that 'we're working on it' when nothing is being done?"

Margaret was an activist appointed to the Governing Board by Governor Lawton Chiles. She didn't care much for the chummy collegial approach to decision-making. She was on the board with a purpose. Her questions and her comments frequently forced debates that dragged on into the night. She was not a favorite among her colleagues on the board.

Still, they give her high marks for intelligence and persistence, if not sociability.

Rich McLean, the deputy executive director of resource regulation, gave a presentation to the board on the steps the District was taking to work with the Authority and its members' utility directors. Rich talked about augmentation as part of a short-term solution. He talked about water-shortage plans and the coordination of a networked water-supply system.

He talked about working with the Keystone Civic Association (from northwestern Hillsborough), the people near Cross Bar, and the people in Land O' Lakes (central Pasco County). He talked about possibilities, and he talked about the future, but he didn't talk about anything that was being done to change things now.

He got one question from board member Jim Cox, a retired phosphate executive. The local governments wanted to augment depressed water bodies. In this case, augmentation would pump groundwater into a surface-water system to keep water levels high enough to keep things alive. Environmentally, as a short-term fix, it can keep stressed systems alive. District staff did not want to use augmentation as a long-term management strategy. They claimed augmentation artificially covered up the environmental impacts of pumping. They also said that augmentation could change the environment because the chemical compositions of groundwater and surface water were different.

"What would you do if an individual came in and asked you for permission to augment their lake?" board member Cox asked McLean.

"If it were for aesthetic purposes, we would say no," Rich said.

Signaling the end of the discussion, Charlies Black, chairman, asked, "Anyone else?" He looked around the board. No one said anything. He looked at the blue cards in his hands. If you want your three minutes before the board, you must fill out a blue card.

"Steve Monsees?" he asked, hesitating a little. The crowd stirred, waiting for a response. Steve walked to the podium in the front of the room, facing the board members. Papers rustled over the exaggerated sound of the microphone as he took a moment to prepare his materials.

He accidentally bumped the microphone. It made a loud thunk in the quiet boardroom.

"Thank you, sir. Thank you for this opportunity to speak to you. I'm a little nervous, but I'll get over it, I guess," Steve said, his voice breaking just a bit. "I'm Steve Monsees, and I address you as a private citizen and a property owner, as the president of the Shady Acres Homeowners Association, and as the vice-chair of the Coalition of Lakes Associations." He began with a recitation of the growing strength of their organizations and then said:

> The purpose of my comments today is simple—they are to inform you of the devastation caused by present water-pumping practices at the Cross Bar Ranch wellfield and to implore you to make the necessary changes to return our private property to its natural state in the shortest period possible . . . I am not complaining to you today of lowered lake water levels—but of the total and complete destruction of all water resources in our community. There is not a parallel in the recorded history of this area, under any drought condition, that approaches the totality of this destruction. All surface water is gone. All wetlands and marshes are gone. Most wildlife has disappeared. The fish and the alligators are gone, and now even the trees are dying. No man or woman or government has this right . . . I have not attempted to provide you with mountains of statistical data to support my comments, as this data is available from your own staff, who I believe support and share my concerns . . . We ask, we implore, we demand that you do what is right and provide the protection of government to us . . . There has not been a day during the past two years that I have not felt violated and angry because of the loss of our lakes, ponds, wetlands, and wildlife. Please restore what is lawfully and rightfully ours. Are there any questions?

The room was silent.

He spoke for six minutes. They ignored the buzzer indicating the end of three minutes and let him finish. When his presentation was over, Steve asked the board if they wanted to see a report on the dried lakes Channel 13 had produced the previous October. Margaret Sistrunk spoke up immediately. "I do." Others mumbled a general assent. As Steve and staff moved to put the tape in place, Executive Director Peter Hubbell covered the microphone and whispered to Linda Sullivan, director of boards and executive support, standing behind him. What he said can't be heard at all. He turned back to the dais and whispered to the chairman, hand still over the microphone and barely audible in the recording of the meeting. "I told her to go get Honey," he said.

***

Linda Sullivan came into my office. I don't remember that she had me go to the boardroom. I don't remember the boardroom at all that day. I do remember Linda trying to explain what had happened and why it was important. I had been at the District less than two months.

Since Pete called, I probably went. Likely, it was all over by the time I got there. I distinctly remember the sense that something big had happened in that room. I asked Steve Haag, the media-relations specialist, to find out what was up and write a news release. Since that wasn't standard for the District, I feel confident that no one told me to do it.

Why did I order a news release? I don't know. I tasked Steve Haag to produce one each day—to show people we were doing something. Public support of the agency was 23 percent, so I might have been trying to create positive visibility. I remember saying to Steve, "You have to paint a word picture. In their minds, media have to see what the board saw today." When Steve brought his first draft, he had his word pictures. I don't remember the context; I just remember the vivid word picture that stays with me today: "turtles upside-down baking in the sun where once there was a lake."

"My God, Steve," I said. "We're the government. We can't say that." It was a notion that I got over rather quickly. In short order, we would say that and much more.

There is no record of the release or that it was ever sent. The document never made it to central files. I went back to managing the investigative reporter, but the District and water suppliers in the Tampa Bay region would never be the same after Steve Monsees's presentation.

Executive Director Pete Hubbell recalls watching Steve Monsees's presentation and thinking to himself, "Man, that sucks." Pete was thinking that he'd never seen Monsees's pond. He had been at the District for years, starting as a staff hydrogeologist. "I had been on most of the water bodies in the District, and it just didn't look familiar to me, so I just didn't have a feel," he said. "But you could tell it was a very personal thing for him. It struck me more on a personal level than a professional one."

After Steve finished, the executive addressed the board. "Many times, the folks that we continually ask to conserve don't necessarily associate overuse with environmental damage. I think that part of our future campaign, our future education efforts, must include the pictures and some of the video we saw here today, that there is a direct linkage between overuse and environmental damage. We've got to get that message out."

Margaret Sistrunk asked the other board members, "If someone walks into a room and shoots one of three people and the other two run around in circles yelling, 'You call 911' while the other says, 'Did you call 911?' and neither ever make the call, and the guy who was shot dies . . . who is responsible for the death? Is it the guy who pulled the trigger or the two who did nothing?"

Board member Roy Harrell watched Monsees's presentation closely. But he was even more tuned in to the board. When he heard their comments, he realized something big was happening. Roy, like Pete, had good political instincts and could tell that this was a volatile situation. When it came to his turn to speak, he pleaded with the board to stay with the deliberate process they had begun the year before. Roy was a lawyer and preferred a stepwise approach to issue resolution.

"What you've had before this is two agencies for the first time coming together not to battle but to work together to reach short-term, middle-term, and long-term goals," Roy told the board. "If we don't have a process, we'll go litigate, but we'll never get the bottom line of solving the problem that you see. We can see the pictures and see the problem. The question is, how do you correct it? And how do you pay for it? This is a collective problem. I can relate. I live on the water, and I can't imagine living on permanent low tide, and basically, that's what you showed us," Roy said, looking toward Monsees in the back of the room. "But there's something else. We [the Governing Board] are pledged to protect the environment and if, in fact, pumpage has done that ... we're not protecting the environment."

Randy Barthle's blue card was next. Dressed in his jeans and work shirt, Randy appeared as what he was—a working rancher. His ranch shared a property line with the Cross Bar wellfield. "As long as fifteen years ago, my father and others tried to warn this board what would happen, and we were told it was not going to be a problem. We had approximately 1,500 acres of lakes on our property. We now have 35 acres. Big Fish was a 300-acre lake. The Florida record for largemouth bass was caught out of that lake. It's been dry for four and a half years. The cypress swamps are drying and blowing over. Your monitor wells go ten feet down, and there's no moisture. I'm encouraged by your comments today. I hope now that things have turned around. Mr. Harrell, in response to your question 'Who's going to pay?' I don't think that's difficult. Who's using the water needs to pay. I am encouraged, but I won't be happy until I see something happen."

Becky Ayech, a longtime activist from Sarasota County, was a regular at Governing Board meetings. She had a hard time speaking, given her anger and tears, fighting for control of her heart and mind.

"This is probably the only time you've ever seen me lose my composure . . . please give me a couple of seconds to collect myself," she said. Then she began. "Every month for the last six years, I have come here to participate in the process. For six years, every month, two days per month, I take off work and drive from Sarasota to Brooksville to come and be told that you are protecting the environment. You're not. You're not. It's sickening. You're not. You can procrastinate if you want, and you can make the finest speeches that you want. I have heard them for six years . . . but you must get on with it.

"I have watched this board, not you people, but historically, this board has ignored the real-world comments of the people who've been before you. You have stacked your experts against them, and they are belittled as though they are not important. They're very important. You can't discuss this anymore. Are you going to do what you're supposed to do? I always thought you were appointed so that you could have the guts to do what you must do. Many of you, when you came on this board, said, 'I can't believe we're still issuing permits.' Well, guess what? Just yesterday, you issued permits."

When Becky was through, there was little more to say. The chair asked Margaret, "Do you have anything to add?"

"I think Becky said it all. This is insane," Margaret observed, and her tone reflected her disdain.

The chairman said, "OK, we'll move on to the next item."

"See? We move on without doing anything," Margaret mumbled under her breath, off the microphone but audible to the audience. Their inaction was exactly what she expected.

The chairman called for a committee report but was interrupted by board member Jim Cox. Jim Cox was the retired phosphate executive. He often insisted that half the water problem was due to the improperly low price of a limited resource.

"Mr. Chairman?" he said, his gravelly voice and southern accent showing deference. He was shaking his head, thick white mane moving left and right. "I just cannot agree that after hearing the heartfelt, eloquent, and well-presented information that we've heard . . . and the gentleman with the slides had about as much impact as any speaker I've ever heard . . . I can't believe that we're just going to move on."

With that, Cox touched off a board discussion that lasted for several hours. "Let's bypass the rigidities," said Cox. "What are we going to do about the problem?"

In the end, the board decided that it needed to do *something*. Staff was directed to prepare for a special meeting in mid-February and to bring some alternatives. Among those alternatives, the board wanted to know the minimum amount of water needed for public health and safety. The task was enormous, and the staff had just three weeks to get it done.

Had it not been for Jim Cox, the board might well have moved on as they had many times before. But this time, they heard what people said. They saw the real consequences of living near a wellfield, and they were ready to do something. They just didn't know what.

Joe Davis didn't speak, but he was thinking, "This needs to be accelerated." As a multigenerational Floridian whose family holdings included substantial orange groves and real estate, Joe understood the tension between needing water for homes and businesses and protecting the water source. The pictures Steve Monsees showed the board had a dramatic and personal impact. "Whatever we were doing, we needed to get on with it," Joe thought. "This wasn't a normal, cyclical dry period, and we need to move quickly to try to bring some relief."

"But we weren't ready," said Roy Harrell. "I was an advocate for change. And we were in a situation where we were trying for a measured response. The board was learning about the problem, and Monsees' presentation moved us from the far right to the far left. He talked about our responsibility, and it just hit the board members, most of them, hard. That started the confrontation. And we didn't have enough facts."

Assistant Executive Director Mark Farrell sat at a table in front of and facing the board, the audience to his back. He was next to the presenters but looking directly at the board. "It was extremely moving. Extremely moving," he said. "Steve Monsees spoke the words that a lot of staff wanted to speak but couldn't. We were working toward it—an admission that the agency had done wrong—that we had allowed too much water to be taken from Cross Bar wellfield—but who would say that to the board? He laid it out in a very articulate way, very direct. It may have been just the sense or the mood, whatever, but it had a tremendous impact. I think a lot of people said under their breath, 'Yeah. This is what we need.' And you could see it in the board. It hit a chord, the way he delivered it. He hit a chord with everybody at the same time. And this board just grabbed on and said, 'We're going to do something about this.' Little did we know how far it would go."

While Steve Monsees was presenting, St. Petersburg utility director Bill Johnson listened, stunned. As a rule, utility directors have a very clear perspective. They are responsible for making sure that their customers, and their community, have safe water to drink every single day. They are responsible for ensuring that when a tap is opened, water comes out. They take pride in knowing that their customers can take their service for granted. Bill Johnson was no different. Two things that utility directors don't want to deal with are price increases and unhappy customers.

When Bill took over as the director for St. Petersburg's utility, he'd already spent many years working for Hillsborough County's water department. He was amazed at the excellent planning that went into securing water supplies for St. Petersburg. In the past, St. Petersburg had some environmental problems, which Bill had worked hard—and worked with the District—to correct. But Steve Monsees didn't live near St. Petersburg's water-production facilities. He lived in central Pasco County, less than a mile from the water-supply facilities owned by Pinellas County. The situation was awful. But it wasn't Bill's problem.

"How come we didn't know this was going on?" he asked himself. "Why hasn't the District told us? Why hasn't the Water Supply Authority told us? I don't know what we could have done about it. It looked like drought to me, but we didn't even know." Johnson didn't see Steve as passionate or personal. He heard a factual presentation.

"If there was one turning point in everything, it was Mr. Monsees's presentation," Johnson observed. "The board was just awestruck." What Bill didn't fully realize was that Steve Monsees's problem was about to become his. The thirty-day clock was restarted each time a question was asked or answered in the permitting process. That is, if there are no questions from the applicant or the District, the permit must be reissued. If either party has a question, the clock restarts.

Since 1992, the District had asked questions and St. Petersburg had answered them. If either side was going to take longer to compile or review an answer, a waiver of time had to be granted. The day before the Governing Board meeting, Bill had requested and was given another extension. Had that extension never been requested, the District might have been forced to issue the St. Petersburg permits or demonstrate why they didn't meet permitting criteria. As a strategy to avoid confrontation, extensions were routinely requested and granted by both sides. That last extension was a decision that would haunt Bill in the months and years to come.

Think of the different perspectives of each of these people. Each felt the power of Steve Monsees's presentation in their own way, through their own lens. Think about the actors. John Parker, the District staff member who would have been in trouble for helping with the presentation and not warning executives. Margaret Sistrunk, who pushed but nearly let the item drop when a retired phosphate executive stepped in to call attention to board action viewed through the revelation of Steve Monsees's presentation. And Steve Monsees, the lifelong leader, a hero for his nation now demanding relief from the only people who could provide it. Steve Monsees, who met a woman at another meeting who told him, "You. You are the one that can make them listen." That woman was Eileen Hart. Each of these people acted on personal conviction trying to find the right thing to do. While Steve Monsees delivered the message, many other people held him up and made it possible.

"God calls us to action," said water activist Eileen Hart, reflecting on Steve Monsees's presentation. All sides considered Eileen among the most moderate of the activists. "I can't tell you how many times I've desperately needed some piece of information, some tiny little thing that I just couldn't find, that I need to link to something else . . . and the phone will literally ring."

Tears welled in her eyes. "There are mornings that I wake up and say, 'Lord, I can't do this today,' and the phone will ring, and it's someone with the information I need. It's like God is saying, 'You can't give up. You must finish this; you can do this,'" and her eyes spill over with the frustration of too many government meetings, too many demands, too long a fight. "You just do it until you lose," she said with finality as she wiped her eyes. "I am not giving up."

In the coming months, God—and his helpers—would place a lot of calls.

Pasco County lake before drought and pumping.

Pasco County lake during drought and pumping

Chapter 6
The District Changes Direction

Steve Monsees's appeal to the Governing Board set in motion events that no one saw coming. "I saw more newspaper stories," he said. "And it seemed like there was more activity, but I can't recall anything especially different."

Others who were watching both board and staff reactions saw things differently. With regularity, interested local governments had staff monitor District meetings. The monitoring was meant to be a kind of early warning system so that if something happened or was about to happen, there would be information beyond media reports.

"When our people came back from the board meeting and told us what had happened, we were surprised," said Ed de la Parte, legal counsel for the Authority. "In the past, when the Governing Board heard this kind of thing, they tried to avoid confrontation. They were not going to let this escalate into a battle with the Authority because we're partners. The Authority was not the wrongdoer, the evildoer, and then our people reported that they were going to have a special meeting and wanted to know the quantity of water for health and safety."

To de la Parte, the board's action was a complete surprise. Identifying a "health and safety level" screamed permit cuts.

On February 7, 1994, St. Petersburg councillor Paul Yingst, who also sat on the Authority board, distributed a memo to the other St. Petersburg Council members. In part, he wrote:

> On January 26, 1994, the Southwest Florida Water Management District Board of Directors ... initiated emergency action after a dramatic presentation by three members of the public. Their action occurred immediately following the staff's report on cooperative planning ... The input from the public had to do with the severe lowering of lake levels and, in some cases, the complete

> disappearance of what once were flourishing lakes in northern Pasco, next to the Water Authority's Cross Bar Wellfield ... Transcripts of that meeting indicate that a rather frustrated board took emotional rather than deliberative action. While this kind of board action does not seem too unreasonable during a water crisis, the conversation of board members tended to expand to include Cosme and Section 21 . . . We are concerned about hasty action with board application that might unreasonably impact the City's wellfields.

Cosme Odessa and Section 21 were wellfields owned by St. Petersburg and jointly managed with the Authority. Yingst offered a resolution for the city council to pass. He wanted it delivered to the District board to address their concerns.

Though the District staff began a massive data-collection and data-analysis effort back in 1988, by 1994, their recommendations hadn't yet been prepared. A program was underway to help board members understand the basic concepts—the depth and breadth of the problem—but the report wasn't ready. In the comments that followed Monsees's presentation, Executive Director Hubbell told the board, "The information on the WRAP [Water Resources Assessment Project] will be ready this summer." Monsees's presentation was made in late January, but the board clearly wanted answers immediately. That's why they ordered staff to hold a special meeting before their next regular board meeting.

"It just hit us," said Assistant Executive Director Mark Farrell. "It was like whoosh, and we were thrust headlong into it. And this board was looking for mistakes; they wanted to see if Pete had a backbone. We were waiting to be told, 'Do your job,' and the staff was ready."

The special meeting was scheduled for mid-February at a hotel in Tampa. The District booked a large ballroom. They were expecting a huge crowd, and about three hundred people showed up.

Today They Call It Optics

By mid-February, I'd been at the District for about two months. I noticed that our media coverage was improving with each definitive action the board took, and I was trying to do more direct outreach with the media and with citizen groups. I was naturally sympathetic to both, so our relationships were improving. I was a blank slate with no real animosity toward the Authority but lots of sympathy for the residents.

I was standing in the entryway of the Tampa office when one of the Authority staff members came in. It was Dave Bracciano, the conservation-program coordinator. I said hello, and we began to chat. The only thing I remember from our conversation is his big, toothy smile, his condescending attitude, and his final words: "We have something special planned for you today."

I knew from his attitude that whatever it was, I wouldn't like it.

I was right.

The auditorium was set theater style with chairs arranged facing in one direction, toward tables in the front, where the Governing Board would sit. It was tight, with chairs touching. The Governing Board and the executive director were positioned to the front and left. The tools for public speaking—a podium, overhead, and screen—were placed so that everyone could see.

People began making their way to seats well ahead of the scheduled start time. The room slowly filled with folks wearing green vests sporting a sprinkler head character embroidered on the back with the words "Mr. Green Thumb."

The green-vested Mr. Green Thumbers came to the meeting to represent the green industry, the people who make their living installing automatic sprinkler systems, new sod, landscaping, and the like. They were the "something special" for me and the District from Authority conservation coordinator Dave Bracciano. He had alerted them to the meeting and told them the District planned to enact stronger irrigation restrictions. No one know what the board might do or what the staff might recommend. The green-industry folks wanted to share their thoughts on the subject, and they were very loud.

Presentations were made by the Authority, some of its member governments, and a handful of others. The Authority and the member governments presented their case to stay the course. Others were for change. The people in the green vests booed the Governing Board as they listened to, questioned, and commented on the presentations.

St. Petersburg mayor David Fischer presented the city council's resolution. In its first public comment on this issue and the resolution, St. Petersburg acknowledged the regional nature of the problem and its concern for the lowered lake levels in other communities. The resolution listed all of St. Petersburg's accomplishments in conservation over the years and concluded with facts about expenditures and actions to improve the wellfields and reduce use. St. Petersburg's resolution was the first step to put their facts on the table. Mayor Fischer didn't know it at the time, but he would soon replace Yingst on the Authority board as St. Petersburg's representative.

"All of it accelerated so fast," the mayor said. He knew this much: "This would be a war that nobody could win."

The District staff liked Mayor Dave. He was a good friend of board member Roy Harrell, and everyone knew the mayor was as good as his word. A financial expert by training and experience, Mayor Fischer was not a man of many words, and he was unflappable in meetings. He doesn't have that gladhanding quality that many elected officials do. People who met him got the sense that he was the real deal—someone who sought elective office because he had something to give, not something he wanted to take.

In the middle of the meeting, I stepped outside to arrange an interview between Sonny Vergara, general manager of the Peace River Manasota Water Supply Authority, and Kevin Kalwary, the investigative TV reporter who had been working for several months on what was going to be a very negative story. I'd badgered the reporter to give our supporters a chance to say something. He relented based on a list of seven names provided by the District executive director; I'd called them all to find someone to speak publicly for us. Only Vergara agreed, having been a District employee back in the 1970s and later an executive director of the St. Johns District.

During the interview, Vergara confronted Kevin. He criticized the media in general for focusing only on the negative side of issues and failing to recognize the importance of the District's job. Vergara got in a couple of good soundbites, and I felt like we had at least tried.

"Did you see his hands shaking?" Sonny asked me. He was glad to help the District. I liked Sonny. He had a low-key personality and a warm smile. He seemed like a standup guy, and clearly, the District didn't have too many of those on its side.

After the interview, I slipped back into the meeting, where the presentations and discussions continued. The meeting lasted for seven hours. Everyone was exhausted. Following my new communication policy, a news release was written and distributed. I used a payphone to make the media calls, then got Executive Director Peter Hubbell to come and talk to reporters. This kind of outreach was new to everyone. Cell phones were neither reliable nor available to government employees at the time. Hubbell gamely stood with me at the payphone as I called key reporters.

On the drive home, Dave Bracciano's comment rang in my ears: "I've got something special planned for you." I thought, "When I get on my feet, you are going to be sorry. Very, very sorry." It was personal now. Nobody screams at my board for seven hours. I didn't like Bracciano, and I didn't even know him.

At their February meeting, the board directed staff to find the minimum amount of water needed for health and safety. The board members thought if they had that answer, they might be better equipped to determine where water savings might be made. The staff didn't have an answer and, expecting a legal challenge, needed to carefully construct whatever methodology they decided to use.

"There was no track record, no paper trail for this action," Ed de la Parte said. "There was nothing in writing saying 'Gee whiz, guys, you need to develop some alternative water supplies.' Nothing." To the Authority's lawyer, it seemed that "by hook or crook, the District identified us as the bad guys and threatened to shut us down within the next couple of months."

But de la Parte's comments are arguable. In 1992, when he was handling the review of the St. Petersburg permits, de la Parte encouraged his clients to deny the District some requested information—it was the same thing Pinellas had done in 1972 as the District was establishing its new regulatory program. It was a move St. Petersburg later regretted.

In 1993, when the District was working on a water-shortage plan with the Authority, the member governments, and utility directors, there was little cooperation. Any suggestion from the District that new, alternative supplies were needed was met with the same response as requests for information on the St. Petersburg permits. Inside the Authority, any attempt to develop water supplies that were not from present sources or new groundwater failed. On the one hand, the absence of paperwork suggests that the District was trying to work through the problem without creating embarrassing and inflammatory documentation.

On the other hand, it could mean that the District was playing a bureaucratic game of keep-away with St. Petersburg's permits. In fact, since St. Petersburg's permits had been under review for close to two years, and they were issued for a ten-year period, it was clear that something wasn't right. Perhaps the vigor of the District's response was a surprise, but it wasn't surprising that the District believed there were serious environmental problems. If there were no concerns, the permits would have been issued.

Between March and May 1994, every attempt the District made to implement even small changes was met with litigation. At first, the District offered water-shortage orders. A water-shortage order is an administrative move to order local governments to take specific action to reduce water demand. District staff drew heavily from the water-shortage plans they'd been developing cooperatively with the Authority and its member governments. The plans were developed well before Steve Monsees's presentation. In a fact sheet dated March 1, 1994, the District made these claims:

> The Water Shortage Order prepared by the Southwest Florida Water Management District staff calls for the West Coast Regional Water Supply Authority and Pinellas County to take steps to reduce the environmental damage caused by over-pumping from the wellfields.
>
> It explains who the parties are and has five bullets explaining what's in the order, including asking for increased enforcement of existing watering restrictions. In conclusion, the fact sheet adds:
>
> The District Water Shortage Order is asking little more than to have the West Coast Regional Water Supply Authority comply with their existing permits, go forward with alternative sources as proposed in their own plans (1986), take some conservation measures to reduce demand and help inform the public about the environmental impacts of taking too much water from the ecosystem. The staff is seeking approval from the Governing Board to proceed with enforcing environmental permit conditions, which have already been violated.

Those working with the District on the water-shortage plan probably felt that portions of it were now being used as a weapon against them. Even with the legal challenges mounted by the Authority, Pinellas, and St. Petersburg, the residents and the water activists were pleased that the District was doing something. To them, the District was exercising its authority, doing its job. They didn't know where it was heading, but they preferred it to nothing.

The Ladies of the Lakes

Between January and June 1994, District staff forged a strong bond with water activists. Eileen Hart, Judy Williams, and others formed a core group of primary contacts. They didn't hesitate to call day or night to request copies of reports, background information, briefings, or whatever else they needed. I didn't hesitate to provide it.

"The Ladies of the Lakes"—as they would come to be called in private by St. Petersburg and Pinellas representatives—could say and do things for the District that I could not. They attended the meetings that I didn't. They were victims. They had a right to be heard, and the media wanted to hear their stories. Together, we were working to right something terribly wrong. With their help, the District was able to largely sidestep its past mistakes—issuing the permits that caused the problem—or at least to own those mistakes but be forgiven. That forgiveness allowed the media to forgive us. Forgiveness let us move beyond blame and into solutions. But St. Petersburg, Pinellas, and the Authority were still trying to blame the District. It was a losing proposition for them, publicly and politically.

"Watching what was going on, I had the feeling that someone was helping them," said water lawyer Doug Manson. "Someone was getting them organized and helping them be effective in their presentations. This groundswell didn't just happen on its own."

In fact, the groundswell did happen on its own, though the District offered help and support, which strengthened and sustained it. In exchange, the District asked for support to change the policy, turn things around, and help get the environmental system back in balance with the water-supply system. Before this shift, the only ones talking to the Governing Board and getting most of the press were the guys with the permits, Manson observed.

"Previously, there was no organized environmental movement to speak of," Manson said. "Now, all of a sudden, there's another voice in the mix, and it was there consistently throughout the process, and ultimately, the politicians joined in that voice because it was popular, and things just snowballed."

De la Parte agreed. "I think the District was very effective in turning public opinion against the Authority," he said. "Things became more personal; people were demonized."

But then again, utility directors, spokespersons, and lawyers acting on behalf of their clients or their organizations refused to acknowledge a problem. Some of them appeared to show no sympathy for people who were impacted or for the environment in general. Representatives of St. Petersburg, Pinellas, and the Authority continued to deny there was a problem and insisted that if there was a problem, it wasn't their fault, or if it was their fault, pumping was only a small part of the problem. And anyway, it was the District that gave permission to pump.

To the District, it seemed like the Authority and its members were splitting hairs. As early as 1968, the US Geological Survey had linked wellfield pumping to impacts; it just wasn't sufficient to stand up in court. Denials of responsibility from St. Petersburg, Pinellas, and the Authority made them seem mean-spirited. The District, by comparison, was trying to help people. Contrasted with the District's apparent admission that something was wrong and needed to be fixed, Pinellas, St. Petersburg, and the Authority demonized themselves. The District and the residents simply helped publicize it.

The Undertow: Litigation, Negotiation, and Frustration

"There was a hardness about our board, a hardness in attitude," board member Harrell observed. The survivor of several changes in governors and their different appointees, he had served through various Governing Board leaders. "Our board had something hard to do. It was clear this wasn't going to end up in a negotiated settlement."

The next few months were filled with litigation, negotiation, and tremendous frustration. The District decided to use its authority to issue water-shortage orders. If challenged, water-shortage orders must be upheld by the court before they can be enforced. As soon as Water Shortage Order 94-12 was proposed, St. Petersburg, Pinellas, and the Authority filed suit to stall it. Water-shortage orders might set targets for additional conservation and require searching for system leaks, public service announcements, and other interim steps. Local governments can order a reduction in or prohibition of lawn watering, car washing, and other outdoor uses. In this case, they decided there would be no changes.

The following month, District staff proposed Water Shortage Order 94-13. Immediately and jointly, the Authority, Pinellas, and St. Petersburg filed suit to prevent it from being implemented. The District was effectively stopped. The board ordered negotiation. Through March and April, negotiations dragged on. Summer rains, when they come, usually begin the second or third week of June. That was an awfully long way off, given the environmental conditions. Things were bad. Rainfall was scant, and everyone's frustration was growing.

For all the political storms raging around the District, a technical storm was raging inside. On one side of the issue was a group that was sure the lack of rainfall was the primary problem with lakes, wellfields, and other water bodies. On the other side was a group that believed the impact of rain (or lack of it) could be segregated from the effects of wellfield pumping. Since the Governing Board had decided that it was time to do something, the staff needed to present them with a consistent technical rationale. Florida law has ensured that the Governing Board consists of people from all walks of life and many professions. It was the staff's job to help them understand the problem. They are unpaid citizen-servants.

The political implications and results of this staff debate had serious consequences. The Governing Board would be asked to make policy and then defend it in court. If the staff decided that the lack of rainfall was the primary reason for environmental impacts on lakes and wetlands, the result would be one kind of management plan and policy. It would be very much like the kind of plan the District had been working on with the Authority since the previous fall.

A decision that wellfield pumping caused severe environmental impacts would lead to another kind of management plan. The assertion that the wellfields of St. Petersburg, Pinellas, and the Authority were the cause of environmental damage would surely divide the agencies like never before. The District needed to resolve these professional differences and present a united front to the board. It was left to Assistant Executive Director Mark Farrell to do that. According to District data, May 1994 was the driest month in a hundred years. And May is the driest month most years. Environmental conditions were as bad as they had ever been.

"The staff was arguing," Mark recalled. "It seemed that there were legitimate points on both sides. We had small workshops where we tried to work out the issues but couldn't seem to reach a shared decision. We needed to resolve it and move forward with developing policy alternatives for the board. I don't remember exactly why, but we decided to have formal presentations, invite the involved technicians, and let everyone make their best case." Mark Stewart, a University of South Florida geology professor, acted as arbitrator. Under contract to the District for several projects, Stewart was also the major professor for some of the District's senior technicians as they pursued advanced degrees.

It wasn't a contentious meeting but rather a scientific debate. John Parker presented for those who thought pumpage was the primary problem. Dave Moore, deputy executive director for resource management, presented for advocates of the rainfall claim. Stewart supervised each presenter's master's thesis. Moore's argument centered on the Cross Bar wellfield. Parker's was on the relationship between the surficial aquifer and the Floridan.

According to the *Water Atlas of Florida,* the Floridan Aquifer is the largest and most used drinking water source in the Southwest District. Withdrawals from the surficial aquifer are usually smaller, except where there is considerable agriculture. Both men had the education and experience to back up their positions.

Stewart recalled the meeting clearly. "The catalyst for this was Cross Bar Ranch and Bob Perry's [District staff member] statistical modeling that he felt showed that significant effects of pumping could be distinguished far beyond the wellfield boundaries," Stewart said. "Dave gave a lengthy presentation based on x-y plots of well hydrographs [water level in a well plotted against time]. He would lay one plot from a well in the wellfield on top of another for a well away from the wellfield and declare that they were the same and, therefore, the wellfield wasn't affecting water levels beyond its northern boundary."

Parker, Perry, and Stewart told Dave that he couldn't just lay water-level plots over each other and draw the conclusion that the water-level variations were from rainfall alone.

"Dave did seem to have a big investment in proving that Cross Bar wasn't causing any problems, although he didn't do anything rigorous to back up his position," Stewart said. "I don't remember exactly what I told him, but I may have said something equivalent to the quote attributed to me."

Someone in the meeting quoted Stewart saying, "It's just incredibly bad science, Dave."

In the end, the staff felt overwhelmingly that sufficient data existed to link groundwater pumping to lake, pond, and wetland impacts. The questions were how strong the evidence was in this specific case and whether it would be enough to convince a hearing officer or a judge. Many District technicians had been ready for years to press this position inside and outside the agency. What they needed they finally got: a Governing Board willing to listen and act on it.

WAR: Words, Anger, Resentment

While the staff sorted through the technical issues, the Governing Board enjoyed some of the best press it'd ever received. As positive coverage increased, the board grew bolder. Internally, the staff were concerned. After years of negative press, all the positive media coverage might encourage them to go too far. But how far was that? What would it take to convince the Authority, Pinellas, and St. Petersburg to resolve the issue? Could negative media coverage be enough to force a settlement?

While that battle was raging, the District was still reviewing St. Petersburg's permits. St. Pete utility director Bill Johnson fumed that the damage wasn't caused by his wellfields but by Cross Bar—Pinellas County's wellfield. The District didn't see it that way.

The District saw a system of seven wellfields that, if managed properly, could be rotated so that if one area was stressed, it could be rested and another area pumped. Doing that, District staff believed that a better management plan and a better environmental outcome were possible, at least until new water sources could be developed. The best answer was new water sources. But local governments were not interested in new water-supply investments.

St. Petersburg had great leadership over the years, and its water cost it about twenty-two cents per thousand gallons to produce. It sold that water to residents for more than a dollar. St. Petersburg's water-supply infrastructure was long since paid for, and thanks to conservation, its water usage was back to 1970s levels. Well situated for the future, it had no interest in spending money to develop costly new water supplies.

The same was true for Pinellas County. New water sources would cost upward of two to four dollars per thousand gallons (maybe more), and an integrated system would force utility directors to treat their water differently so that the sources could be blended. That was an additional expense. The fact that present water supplies were in someone else's backyard was immaterial to the Authority, St. Petersburg, and Pinellas. They maintained that drought was causing the problem. This was not an issue of pumping, and if it was, the District had issued the permits. They had done nothing wrong, they insisted.

The District decided that the problem was pumping, and the staff thought they could prove it. The Authority, St. Petersburg, and Pinellas believed they could prove otherwise. In truth, Pinellas expected the District to back down. Few believed matters would go as far as they did.

Still, by March 1994, every local government and even some of the activists retained counsel and prepared for war. There were in-house lawyers, outside counsel, general counsel, and experts on all sides—all paid for with public dollars—except for the residents and activists. For example, a resident of St. Petersburg who pays for city water and ad valorem property taxes also paid for (1) St. Petersburg staff lawyers, (2) outside counsel for St. Petersburg, (3) an Authority staff lawyer, (4) outside counsel for the Authority, (5) District staff lawyers, and (6) District outside counsel. The Authority sometimes had inside and outside counsel and sometimes only one or the other. Since each member government pays to be part of the

Authority consortium, St. Petersburg residents were helping pay for those bills. Customers were paying for at least six lawyers on all sides of the case to determine what each was supposed to do—deliver water at an affordable price without damaging the environment.

St. Petersburg residents continued to pay lawyers on all sides for at least four years. In the spring and summer of 1994, there was no agreement on what policy best served the public interest. But in two more years, the public and the media became very interested in the high cost of continued disagreement.

Legal maneuvering dragged on while the District waged a public war of words. Never had the District set out to gain public support. When the media covered conflicts among local governments, the Authority, and the District, it characterized them as squabbling. Although the intergovernmental conflict wasn't dismissed as unimportant, the use of terms like "bickering" and "squabbling" trivialized the issue.

The people closest to the issues, those who lived near the wellfields, were following the District's latest actions closely and liked what they saw. To St. Petersburg, Pinellas, and the Authority, the District had "turned without signaling." For District staff, there was finally board support to do something tangible about a real problem. The District wanted public support. For the first time, it worried less about what the local governments would think or do and concentrated more on developing public support. It was a direction and a message that worked well.

District executives finally felt they could push their agenda—getting the wellfields back to a sustainable level of pumping. The District had overpermitted the resource, and before those permitted quantities were reached, the negative environmental consequences were obvious. Tens of thousands of wetland acres were devastated, and some would never recover.

Filtering the Message

The District called the problem "overpumping." The Authority, Pinellas, and St. Petersburg claimed that they were not overpumping; they were within their permitted quantities. To them, overpumping meant that they had exceeded the quantities allowed by the District. From the District's perspective, overpumping meant that too much water was being taken from the ground. It was a policy impasse.

The District admitted it had over-permitted the resource—the activists had insisted on the admission—but even so, local governments and the Authority refused to recognize it. The battle over language was a sensitive one. Still, arguments about whether the District overpermitted or the permittees overpumped were mostly lost on the public.

Even a layperson could see the treefall and subsidence, the dry and dusty wetlands, the rain gauges rising from sand and know there was a problem. Subsidence is an indication that water levels are low. It is a condition where the soil has fallen away and left sensitive areas of the root exposed. In a swamp, this condition can leave cypress tree knees unnaturally exposed. When that happens, the roots can no longer sustain the weight of a tree, and it falls. This process occurs naturally. But when natural hydroperiods are unnaturally altered, entire swamps can be devastated. In northwestern Hillsborough and central Pasco Counties, the District estimated that tens of thousands of wetland (swamp) acres were impacted by wellfield pumping.

The District wanted to fix the problem and sought public support to do so. The District needed a popular constituency because the political forces on the other side were powerful. The distinction between overpermitted and overpumped would not be lost on the courts, however. A finding, either way, could determine who would be responsible for paying for the new supplies if and when they were developed.

* * *

I didn't experience the cyclical relationship of the Authority and the District or the years of frustration from residents and activists. I was working in marine-ecology education at Mote Marine Laboratory in Sarasota. I knew very little about freshwater issues. Like the public, I never thought about where my drinking water came from. I saw the challenge of water-shortage orders the way the public did. I couldn't understand why *the other side* refused to acknowledge the problem or even consider that wellfield pumping might have a negative effect. It seemed logical, and District technicians explained how the natural system and the water-supply systems worked so it was understandable. The system just never got to stasis, where drawdown and pumping reset the water levels in a natural system to a lower point. The predicted leveling never occurred. Instead, water levels just kept dropping and dropping, with the water being sucked out of Pasco and Hillsborough Counties and sending it to Pinellas County and St. Petersburg.

It was my job to tell the District's story. I came to the District a blank slate, but my education was rapid (if one-sided), and I didn't like what I saw of Pinellas, St. Petersburg, and the Authority and their denials.

Armed with my outrage, growing knowledge, and alliance with the residents and activists, I had everything I needed. Just a few short months before, in January, when I looked at the news release about Steve Monsees's presentation, I removed "turtles upside-down baking in the sun" as too strong for the government. In a few short weeks, my growing indignation led me to draft copy that made that news release on the turtles look like an invitation to tea.

Board member Joe Davis dubbed part of the problem the "two-newspaper conundrum." The community where the water withdrawal caused damage had one daily newspaper; the community benefiting from the water had another. Until there was shared understanding of the problem in both newspapers and both communities, people could not or would not understand. The water activists felt that the District had always been pushed around by local governments. They were afraid it would happen again. District leadership insisted that they were trying to fix the environment, and many believed them. They wanted to; they needed to.

This moved beyond the pressure to change and the resistance to change. The people involved had entrenched viewpoints based on years of experience. To effect change in policy, people needed to change. People drove the conflict on both sides. Their words, their actions created a context in which resolution was virtually impossible. To change the equation, the people embedded needed to change.

Chapter 7
The Watermark

Steve Monsees started this chain of events in 1994 with complaints about Pinellas County's Cross Bar wellfield. But St. Petersburg's permits were still the only ones under review by 1996. The Cross Bar permit wasn't in yet. The District wanted to have all the permits to evaluate them together to treat them as a system. Changing one or two permits wouldn't do much for a systemic problem.

Joe Davis, now District chair, along with executives Hubbell and Farrell continued to try to work something out with St. Petersburg. Despite their initial elation when water lawyer Doug Manson joined the St. Petersburg litigation team, nothing changed.

"I didn't think that the District was going to starve us out," St. Petersburg mayor Fischer said, "but I thought that they were serious enough to put all of the leverage they could on the wellfields."

St. Petersburg, perhaps tired of having its water-supply permits held for so long, finally refused the District's request for an extension. Since St. Petersburg demanded its permits be issued, the District had two choices: grant them, or deny them. In choosing to issue the permits, though, the District could modify them. It considered this alternative, offering St. Petersburg one-year permits with conditions. District staff also considered some "stepdown" permit conditions, which would dictate successive reductions in pumping over time, allowing St. Petersburg to replace the water it "lost" with new supplies. Ed de la Parte convinced District staff that the levels in the stepdown permits couldn't be defended in court. He would see to it. So, even if St. Petersburg wanted to make a deal, Pinellas made sure it couldn't.

"District staff wasn't competent enough to form a new scientific standard," said Manson.

On the other hand, the District could deny the permits. But denying permits was a risky, draconian step. After all, this wasn't golf course irrigation; this was public supply—drinking and baby-washing water. Could the District take such a step and still get people to believe it wouldn't turn off the flow of water? The District couldn't shut off the water. It's a demand system, meaning when the tap is on, water flows. From a legal perspective, either course was fraught with peril. If it issued modified permits, the legal burden would be on it to prove how and why it could work. If it lost, public support would surely disappear. If it denied the permits, it risked political wrath greater than anything seen thus far. The phony cries of a building moratorium could become very real. The potential economic consequences were significant. But the District couldn't just reissue the permits without changes. It knew those permits were causing environmental harm. It did not want to issue the permits as they were, and it couldn't, according to law. Its mission was to prevent harm.

Roy Harrell followed Joe Davis as board chair. When executive staff told him that the District needed to deny the St. Petersburg permits, he threw them out of his office. Harrell was a friend of Mayor Fischer's. The issue wasn't so much their friendship; it was more that the mayor was a good guy, he was coming around, and everyone knew it. Denying the permits would inflame an already-charged situation. But then again, the mayor had refused another extension and demanded the permits be issued or denied.

"When we went in to see Roy the next time, we took everyone with us," Mark said. "There was a lot of turmoil. We had to come back with something."

***

We sat at a table in the conference room of Roy Harrell's law office. It was a circular room with wood trim and no windows. No distractions outside. Roy sat with his back to the door, the only way out. Although he was his usual affable self, it was clear he was not happy. It was never a happy meeting when District staff, particularly this crowd, migrated to his office. A meeting like this was unprecedented anywhere in Florida. District executives were preparing to ask the chairman of the board to support a decision to deny water-use permits for public supply—permits that had already been issued and reapproved years before.

Seated at the table were executives Pete Hubbell and Mark Farrell, General Counsel Ed Helvenston, lawyers Karen Lloyd and Jim Robinson, and me. Roy did not want to hear what the staff had to say. We were going to make a recommendation to deny the permits.

Pete started with a review of the latest round of negotiations. Former chair Joe Davis and Pete had worked hard to come up with a plan, a permit, that St. Petersburg could agree to, but that failed. The District was on a statutory time clock. We had to do something, and we needed Roy's support to do it.

Roy turned to each of the lawyers, one by one, to ask their opinion. Karen handled rulemaking, so she spoke from that perspective. Jim was the litigator, so he put together the strategy and would try the case. This was going to court. Roy was a lawyer and the longest-serving board member, so he had more than a basic working knowledge of how the process might work and some of the alternatives.

But Roy was a real estate lawyer, so he deferred to the District's legal team. I sat directly across from him. Periodically, he'd glance my way. We always had a special relationship—he said that I was as pigheaded as him. Finally, Roy asked me, "What about you? What do you think?"

"We've come this far," I told him. "I think whatever we do, we better win. You guys find the strongest possible legal position, and I'll figure out a way to explain it."

"Roy, we don't have a choice," Mark told him. "De la Parte has made it impossible to do the stepdown permits, and we can't issue them like they are. If we don't do this now, we'll never get control of this situation again."

Roy didn't like it. He lived and worked in St. Petersburg. He knew, as did District staff, that the city wanted to work things out. But for reasons no one could fully explain, we were at a policy impasse. Reluctantly, he agreed.

"You had better be right," he cautioned. "Denying public-supply permits—it's never been done."

"For us," Mark recalled, "it was better to know. Either we went to court and we won, in which case we'd done our job. Or we lost, in which case we'd appeal. If we were wrong, we were wrong. We needed to know. It was better to know. We needed to bring it to a head."

As we crowded into the elevator on the way downstairs, we were not elated at the "win." There were no high fives for getting Roy to agree. We were quiet. We were thinking about what might come next. The lawyers were thinking about papers to be drafted and filed. Jim Robinson was perhaps thinking about his friendship with Pinellas commissioner Steve Seibert. Soon, District legal actions would preclude them from speaking. Pete and Mark were thinking about briefing board members on the staff recommendation, and I was thinking about the story. How was I going to tell this story?

April Fool's Day

It had been an intense two years. Litigation. Investigations. Politics. Employee firings. Resetting priorities. By April Fool's Day, 1996, things weren't as tough as they once were, but we were in the second year of ongoing stories by an "investigative" broadcaster; this time it was the NBC affiliate. It was no secret his reporting annoyed me. I was frustrated. I tried to keep an in-your-face attitude with him to bolster staff morale.

Somewhere along the line, I decided Mark and Pete were overdue for a good April Fool's joke. I recruited Linda Sullivan, the director of boards and executive support, who got stationery from the Authority, and Sandra Anderson, who was the communications-department manager.

I wrote a letter to myself on Authority stationery from Jerry Maxwell, the general manager. In the memo, I wrote all the usual buzzwords I continued to stress inside the District. In the letter, Jerry encouraged me to keep our meeting confidential and then proposed a relationship that would begin with consulting and conclude with a senior-staff appointment at "the proper level" within the Authority. We got the executive secretary, who was the best forger in the District, to sign Jerry's name.

On April 1, I was in the Tampa Service Office. District headquarters is in Brooksville, Hernando County, north of Tampa. The drive from one office to the other was about forty-five minutes, depending on traffic. In my absence, Sandra "found" the letter, opened it, and rushed to Pete's office to show him. She told me later she felt proud of her performance.

"I found this in Honey's mail," she said to Pete breathlessly. Her nervous breathiness probably lent gravity to the document she showed him. "I always go through her mail when she's not here," Sandra told him. "I thought you should know."

Her serious hazel eyes looked directly into his. She handed Pete the letter, and he read it. Sandra reported that Pete went white. "Wait here," he told her. She waited while he went to make a copy of my private correspondence. He returned, thanked her, and told her to return the letter. "Don't worry," Pete reassured her. "He was pissed," she giggled into the phone, still flush with excitement. We shared a sneaky laugh at our immense cleverness.

I got into my car and drove up to the Brooksville office. When I got in, I asked for another report from Sandra. Upon reflection, she was a little concerned. Pete was pissed, and Pete was never pissed. Linda Sullivan also came downstairs to report her observations. "He is furious," she said. "How long are you going to let this go on?"

I thought I ought to gauge the situation myself, so I strolled through the executive offices upstairs. As usual, I poked my head into Pete's door and said hi. I got a very curt hello. But someone was in his office.

Hours went by, and I decided Pete shouldn't go home angry. At 5:15, I gathered all my coconspirators and went to his office. His door was ajar. Fritz Musselman, director of land resources, was in the room. "Hey, Pete," I said, with everyone gathered behind the half-shut door and me.

"What do you want?" he hissed, his tone decidedly nasty.

"I just wanted to say hi," I told him.

"I'm busy," he said flatly, turning away. Behind me, I could hear the buzz. Pete simply didn't talk to people like this.

"I just have one thing to say to you," I told him.

"What's that?" he snarled.

"Do you know what day it is?"

He paused for a moment. I could almost hear the gears turning in his head. A beat, then two, passed, his eyes crinkled, and a smile broke as it dawned on him what I had done.

"You bitch," he said. I opened the door, and everyone poured in. "I can't believe you did that to me." We all had a good laugh.

"Does Mark know?"

Later, Pete showed the letter to Mark, indicating my plans to leave and "go to the dark side," as we used to say. Pete had his fun with Mark overnight, then told him it was a joke. Mark called me names, too.

Little did I know the reason my joke seemed so plausible to them was that they were already planning to leave the District and leave me behind.

Deep Denial: No Water Shortage Here

Since January 1996, Pinellas County officials had worked on a public outreach campaign. They redesigned it several times, finalizing what they believed was a soft sell. One thing was sure; it was long awaited by all parties in the water wars.

"I saw a copy of the insert before it went out," said Mayor Fischer. "I called Pinellas commissioner Bob Stewart and said, 'Are you going to do this?' I knew they'd spent a ton of money on it, several hundred thousand dollars, but Bob is a reasonable guy. 'This is going to make you look bad, you really ought to pull this,' the mayor told Stewart, who replied, 'We already spent the money. It's out there.'"

Pinellas County was moving ahead with its PR campaign. "I think," the mayor said, "that solidified my feeling that Pinellas wasn't going to give on this, that they were going to prove that they were right, and they were going to fight everybody to do it, and we're not going to build any new water supplies. It just reinforced my resolve that we had to change course."

The Friday before the tabloid was scheduled for release, someone provided a copy to the District. We spent two days responding bullet by bullet to every item that needed a response and prepared a package of material, which included photocopies of the tabloid, our response, and some additional background information. The kits were sent out for delivery to all fifty-five Governing Board and basin-board members.

The written responses to the tabloid also were sent to more than seven hundred District staff members so that if asked by neighbors, they could respond. Then, the District generally geared up for whatever was to come on Sunday—the day Pinellas planned to have its tabloid distributed in both the *St. Petersburg Times* and *Tampa Tribune* newspapers.

Its newspaper insert hit early Sunday morning.

The first call came from Channel 8, the NBC affiliate. They agreed to meet me at Jumping Gully in Pasco County. I wanted to show them the pictures of Jumping Gully before the Cross Bar pumping; then I wanted them to see Jumping Gully for themselves—after pumping had drained it.

Those were the days before newsrooms—both print and broadcast—were dramatically cut back. Even though there was weekend staffing, which was generally lighter, they sent a crew. When we arrived, there was a boy chipping golf balls in the dry creek. He was about ten years old and could barely remember water in it. His mom was there, and she remembered; she was happy to get on camera and comment. After finishing with Channel 8, I called Channel 28, the ABC affiliate, and invited them to meet me at some dried-out wetlands. They did. I called Channels 44, 10, and 13 in turn. They didn't have crews free, so I drove video footage to each studio. As I left the Channel 10 studios, my last stop, it was dark. I got into my car, and the skies opened, pouring rain. Crossing the Gandy Bridge, I called Mark and Pete and told them to turn on the news to see what happened. When important water decisions were made, it always rained.

We were on every channel.

The rain was coming down so hard I could barely see. I couldn't help but think, *"If the rain continues, it's going to be hard to explain that there's a water shortage."* But we sure needed the rain, and it smelled so good. It felt good, and it was going to keep the wetlands alive—or so I hoped.

On the last page of the Pinellas tabloid, there was what we in the communication business refer to as a call to action. Pinellas had begun to advance a peer-review idea that District staff believed was designed to delay our advancements, stop our progress, and keep us from finishing the job of changing water policy. The call to action encouraged people to call a number in support of scientifically based, binding peer review and claimed that "Swiftmud's Governing Board has spurned the opportunity."

It asked people to "vote" for Swiftmud to reconsider peer review. It was meant to be a positive approach—who wouldn't support peer review if you knew what it was? The District supported peer review and, in fact, had tried to get Pinellas to engage in peer review the year before. But not now, not when staff felt themselves closing in on a resolution. In response, the District distributed a guest column to daily and weekly papers in the tricounty area, with Pete's byline, "Make the call—for the right reasons." It contained some harsh words, but to gain support, the District needed to be clear. The staff were angry about the campaign, and others shared those feelings. Pete's editorial said in part:

> Pinellas County has always used its lawyers to advance its confrontational and selfish water policies. In the past few years, however, as the public has come to better understand the relationship between water supply and environmental damage, Pinellas has had to try new and more desperate tactics . . . If there

> were not so much at risk, Pinellas County's posture could be considered ridiculous. After all, it is Pinellas County officials who have their scientific reports stashed in their lawyer's office. A judge will have to force them to be released.
>
> On the other hand, the District has been making public reports for more than two years . . . Who has something to hide? They are using peer review as a tacky public relations trick and hoping that the public will buy it.

Years later, when I read it, I wondered, What was Pete thinking? Can the government say these things? Should it? Should we have done it? What was I thinking? And yet we had tremendous public support. I believe the clarity of our messages got through to people. Back then, the government didn't say that kind of thing. But we did say that, and we said a lot more. Pete Hubbell, Joe Davis, and all the rest were brave to make this move. It could have backfired. But we had come very far in two years, and we only had a few more months to go, so we took the chance. We asked people to "make the call" and tell Pinellas County officials, *"Tampa Bay can't wait for their expensive lawyers and their advertising campaign. Make the call. We need water, we need to protect the environment, and we need both now."* I have no idea how many calls Pinellas got, but since they never publicized the results, I believe it didn't go their way.

"The PR plan was tacky. We were hurting," said Pinellas County commissioner Steve Seibert. It's a painful memory for him, another low point for the Pinellas County Commission. "It was the frustration—we just couldn't get around the curve to get a message out."

A July 1996 survey conducted by the University of South Florida showed 80 percent support for the District. I was so stunned I made them recheck the numbers. When was the last time a government agency had that kind of support? In three years, we'd moved from roughly 20 percent support to 80 percent despite our very aggressive rhetoric.

In my files, there's a note from someone named Bill Hartmann. On the cover sheet I'd sent him regarding the tabloid, he penned, "Honey, thanks for the informative memo. From my vantage point, it all boils down to all affected parties doing the right thing."

Fishing in a Dry Lake

Pinellas County was getting battered, PR-wise, according to Hillsborough County commissioner Ed Turanchik. Everyone else was trying to make some solution work, and Pinellas was obstructionist.

"We needed to show that we could work together, we were working together," Turanchik said. "First we got a water-supply plan, then we had to figure out a reorganization—governance—because it was broken, and everyone understood that. We had this 'United Nations'–veto circumstance, where one of the superpowers could stop a project by refusing to participate in it financially."

He was talking about Pinellas County. It seemed that whenever a new water-supply project started to gain some momentum, Pinellas would refuse to help pay, and it would stop. The list of aborted projects was long and included the Pasco Rainbow, which was a project to put reclaimed water onto three Pasco wellfields to rehydrate them. It also included the Tampa Water Resource Recovery Project to put reclaimed water back into the natural system as indirect potable. Rehydration and indirect-potable projects pump highly treated wastewater into the natural systems. It migrates through the natural system, getting filtered along the way before making its way back into the water supply. While this is common now, at the time it was innovative.

Water projects in Tampa Bay were never completed, because of Pinellas County's objections. Ironically, St. Petersburg already had a rehydration project for its Section 21 wellfield in northwestern Hillsborough. Each new project was killed by Pinellas County (overtly or covertly) over several years. Pinellas had its water; it had its wellfields. Under Commissioner Rainey's direction, it did not need anything else. If others needed new water, the Pinellas's attitude was "They should go and get it."

Turanchik's position was conflicted. As the chair of the Authority board, he had to supervise official Authority actions to thwart the District's efforts. As a Hillsborough County commissioner, he supported the District.

Other local governments with more open-minded elected representatives had a different outlook. New technologies needed to be considered, they thought. When Pinellas simultaneously killed the Pasco Rainbow project, sued citizens, and allocated $800,000 for a public relations campaign, it was too much. Hillsborough and Pasco got together to tell their side of the story.

Ed Turanchik was both a Hillsborough County commissioner and chair of the Authority. Jim Norman was a Hillsborough County commissioner, and Ed Collins from the Pasco County Commission joined Ed Turanchik in a news conference at Big Fish Lake, the one where the record bass was caught, the one on the Barthle Brothers Ranch. The one that was nearly gone. The one next to the Cross Bar wellfield.

Several of them sat in a landbound boat while Ed Collins stood in the back, looking uncomfortable with one foot in the boat and one on land. The two daily newspapers showed up along with two of five television stations. *The St. Petersburg Times* reported the June 24 event: *"In what might have been the social event of the water-war season, Hillsborough County Commissioners Ed Turanchik and Jim Norman and Pasco Commissioner Ed Collins stood in a boat beached on the dry lakebed that used to be under 15 feet of water. They dropped lures into the surrounding weeds. 'The plan is to cast for bass, and if that doesn't work, we'll dig for trout,' Turanchik said."*

All he snagged was a television soundman.

"Pinellas had their big PR plan, and we skewered it with our news conference," said Turanchik, energized by the theater of it. "Jim Norman, Ed Collins, and I went out to fish in Big Fish Lake, which was dry from Cross Bar wellfield pumping. We got plenty of publicity, and even editorials and cartoons, and Pinellas residents had to be asking themselves, 'Are our lawyers and are our experts telling us the truth?'"

By that time, all the local governments had spent millions on litigation, and Pinellas was still claiming there was no water-supply problem. Before it was over, the cost of the water wars would top $10 million—with not one new drop of water served to the public.

"It wasn't helpful," said Seibert of the staged event. "Much of what we were doing to each other wasn't helpful. We needed to lay down arms and talk through some of this stuff. We were angry; we were lobbing grenades, verbal grenades, on each other, and no one was listening."

Still, Pinellas showed no signs of backing down. The District, Hillsborough, and Pasco moved their joint case forward.

"When Lee Arnold comes into my office and says, 'Where's my water?' I must have an answer," Seibert said. "I didn't have the answer." Lee Arnold was an avid outdoor enthusiast and successful business leader in Florida. His primary business? Development and property management. He was not to be ignored. His business interests bring tremendous economic benefit to the region. Remember the Korean chip factory? Without a stable water supply, development could be at risk. "The fear is, as an elected official, that it will blow up on your watch," Seibert said.

"Pressure was on the situation, not on us directly," Turanchik said, speaking more as a Hillsborough County commissioner than as an Authority board member. "I was overjoyed at the District's position. They were in the right place. They were the regulators, and they were out there trying to get control of the situation. It was a long time coming," he added.

The major changes started with individuals. "I was thinking to myself, 'We might not win this,'" Pinellas commissioner Seibert said. "Even if we win in court, we might not win in the long run. We were taking a beating in the media, and we were getting all our information from one source inside the county. I started to hear things that were different from what our advisers were saying, and I was concerned."

Seibert asked District executive director Peter Hubbell to take him on a helicopter trip of the damaged areas. He saw the damaged wellfields and surrounding areas. Like Mayor Fischer two years earlier, he could see the gaps in his advisers' arguments. It was not so much that they misled as that they looked at the problem narrowly.

In the meantime, Mark Farrell and Roy Harrell were pushing for desalination, while Pinellas utility director Pick Talley vehemently opposed it. The District asked Seibert to find out for himself what state officials said about such a project. Seibert, Farrell, and Harrell headed to Tallahassee, where staff members from the Department of Environmental Protection assured him that a desalination plant was permittable—maybe before new wellfields. "To that point, I'd been given the party line on desalination," Seibert said. "After that, in all good conscience, I couldn't say that discharge brine would be toxic to nearby water bodies."

It was direct information. The District thought its arguments were so strong and persuasive that anyone could be convinced if only they could hear the facts. Their facts.

"We were missing an open back-and-forth debate of these important issues," Seibert said. "We were trying to choose among the experts. Is this one better than that? It was infuriating. I was seeking truth. What I discovered is that science is an art, too." He eventually concluded that the proper answer would come from neither science nor politics alone but from both. He believed there was a lot of scientific guessing going on. That's one of the reasons Pinellas suggested, lobbied for, and pushed for peer review as an alternative to litigation. But the District had called for peer review the year before, and Pinellas had flatly refused.

A reliable version of the truth was out there, but to find it, Pinellas officials needed to accept that pumping was a problem. The District needed to accept that other factors could have impacted the environment. With a court date on the horizon, few wanted to delay long enough to work it out. Everyone could see the future, and it was uncertain.

Do the right thing. That's what most people wanted. But how do you decide what's right and for whom? Mayor Fischer and Commissioner Seibert were the first to seek information outside their usual networks. Both understood that the cost of the litigation went well beyond court. The stakes were high economically and politically. Though they represented communities that had secured water supplies, they began to understand the cost of that security and to question whether continuing as Pinellas and St. Petersburg had was the right thing to do.

Chapter 8
Change Is Constant

On April 1, 1994, Harold Aiken, the controversial general manager of the Authority, resigned. Few people remember Harold fondly. It was one of the occasions when Pinellas commissioner Rainey and District executive director Hubbell agreed. A former District employee, Aiken wasn't one to join in the shared staff fun of the early years. "He never played volleyball at lunch," Peter Hubbell noted. "And he didn't like me."

* * *

On my first day at the District in December 1993, I attended a meeting with Aiken and Hubbell. We sat in the executive conference room where I'd been interviewed for the job. Present were Hubbell, Dave Bracciano (the Authority conservation manager), Harold, and me. Everything was new to me that day, but I do remember Pete saying to Harold, "Water levels are as low as they've ever been. We're not even in the dry season yet, and already it looks like you may break reg levels." Regulatory (reg) levels were set by District staff to monitor the water levels underground. Permittees were not allowed to take any more water when those levels were reached. Typically, local governments would shut down wellfields before violating the standard. If they exceed permitted levels, they can be fined up to $10,000 per day, per violation. But that has never happened. (Reg levels no longer exist. Minimum flows and levels have replaced them.)

"What are you going to do?" Pete asked.

I don't remember what Aiken said, but I do have an impression of it. Aiken gave Pete a kind of "Yeah, yeah. Sure, sure. Whatever" response. He wasn't interested in Pete, and he wasn't interested in the Do Your Part conservation campaign, which I was at their meeting to explain.

As tensions between the District and the Authority escalated, a search for a new Authority general manager began. In the interim, Bruce Kennedy was selected to run the agency. Bruce was a professional engineer and a water-utility manager suddenly thrust into the middle of hardball hydropolitics. Potshots at Kennedy from Pinellas commissioner Rainey punctuated Authority board meetings, while the District publicly vilified the Authority. Kennedy tried to find a way to balance the competing interests and responsibilities and still be himself. He was a nice guy.

On the other hand, Pick Talley, the Pinellas County utility director, like Ed de la Parte and others in this dispute, typically elicited strong negative responses.

While St. Petersburg utility director Bill Johnson was a lifelong public servant, Talley was a former county commissioner, consultant, and public servant. His closest friends said he was an environmentalist. People who didn't like him, people who lived around the wellfields, said he was neither environmentalist nor humanitarian. Like his closest colleagues, Chuck Rainey and Ed de la Parte, Talley was in the policy arena to win. He did not lose gracefully; he was relentless, and his views about Florida's water supplies and policies hadn't changed since the 1970s. He became the head of the Pinellas Water Department in the 1970s and returned to reprise the role in the 1990s after various positions in the public and private sectors.

People either agreed with his policies and positions, or they disliked him with a passion. With him and de la Parte and Rainey, it was personal. People didn't just object to their policies; they attributed malice to the men. District staff would frequently refer to his assertions, as reported in the papers, as "Pick knows." It was juvenile, but just saying it aloud was great fun.

At the time, Pick was a leader among the utility directors. The Authority had a Technical Advisory Committee, which was the way the utility directors from the member governments provided feedback and advice to their board. The group was sometimes used to prevent the development of new water-supply initiatives on technical merit.

Pick Talley was a Hillsborough County commissioner in the 1980s during a scandal. Several of his colleagues went to jail for bribery following an investigation. Talley wore a wire to help law enforcement. Together Rainey, de la Parte, and Talley had led, pushed, and managed the Authority's and Pinellas County's water policies since the late 1970s.

While some deny their coordination, the policies between Pinellas and the Authority were curiously consistent. Talley never sounded sympathetic to the people and environment impacted by wellfield pumping. His total refusal to acknowledge the possibility of negative impacts was arguably part of the problem. And, to those impacted by the consequences, it was personal.

***

Since the special meeting in February 1994, Pinellas had repeatedly suggested that if things were so bad, the District should ban all outdoor irrigation. Talley insisted that the drought was the problem. He denied that wellfield withdrawals were. At the same time, the Authority was calling green-industry representatives to put pressure on the District to leave watering restrictions alone. The show of power from the green industry was meant to intimidate the District, to make its board look weak, to force it to back down. The local governments wanted the District to impose stronger watering restrictions, a move that was unpopular with both the green industry and the public.

In the past, critics said, the District backed down to Pinellas and the Authority. This time, things were different. The Governing Board and the staff had a constituency, and it was growing. They had support, and they were not going to do anything to jeopardize it. The District already had watering restrictions in place, they argued. Those restrictions applied to all sixteen counties the District regulated. Local governments had to decide whether greater restrictions were needed in their areas. The District put it right back on them.

At first, editorial boards criticized the District for complaining about the water-shortage problem and then refusing to act on it, but that position was short-lived. The District got in front of the issue to tell its story by calling reporters and editorial writers and explaining the tactics behind Pinellas's behavior. The District framed that behavior in terms of its impact on water supplies and the environment. I pointed out the hypocrisy of demanding the District add new restrictions while encouraging the green industry to fight them. This pushback was a first for the District.

Finally, on April 16, 1994, a *Tampa Tribune* editorial noted, "The water shortage is real. But the Authority's stand would be more credible if the agency itself were not fighting water restrictions."

St. Petersburg, Pinellas, and the Authority wanted restrictions on homes but not on the utilities that provide them water. They wanted the District to outlaw car washing and lawn watering but not to reduce the amount of water that the utilities took from the source. Their position was untenable. Dishonest. As areas near the wellfields suffered through the driest May in a hundred years, the Authority urged residents to conserve. While litigation and negotiation dragged on, the Governing Board, the District staff, and people concerned about the environment became increasingly frustrated with the Authority, Pinellas, and St. Petersburg for their resistance to the water-shortage orders. The orders were stalled in court until a judge approved them.

The District looked to June rains with mixed emotions. Rainfall would bring some relief to the strained environmental system, but at the same time, it might reduce public pressure to solve the problem. The June rains were disappointing, but still no shared solution could be found. On June 4, 1994, a *Tampa Tribune* headline announced "More Bay Area Wells May Turn Off the Taps." In the article, St. Petersburg utility director Bill Johnson said, "We're the first . . . Unless there's rain, there definitely will be a domino effect." And yet, at the same time, St. Petersburg had joined Pinellas and the Authority in a challenge to two water-shortage orders. At best, it sent a mixed message.

On the one hand, St. Petersburg said, "There's a problem" while on the other, its officials vigorously resisted the District's attempts to manage it. Though newspaper, television, radio, and the limited-at-the-time Internet carried news about the water-supply problem, and public opinion leaned heavily in the District's favor, the Authority, St. Petersburg, and Pinellas were not alone in their resistance to additional conservation efforts. On June 11, 1994, a *Tribune* reporter wrote: "Barbara Meisman of South Pinellas County knows about water conservation—and hates it. She lived in the San Francisco area during the 1970s and was forced, by California's water shortage, to watch every drop she used . . . 'It was the most miserable time of my life,' said Meisman, a retired hairdresser who returned to the St. Petersburg area in 1976. 'It would make me so unhappy—make me the most miserable person in the world—if I had to give up 10 gallons of water a day.'"

That's what the District asked people to do. Just save 10 gallons per day—not enough to affect their quality of life. In the Tampa Bay area, water use was (on average) under 150 gallons per day. Some communities were higher, some lower. Most potable water in urban areas is used for outdoor irrigation, so by reducing irrigation a lot of water can be saved.

A turning point came on June 13, 1994, when Sue Landry of the *St. Petersburg Times* submitted a public-records request. Florida had some of the most robust open-records laws in the US, covering everything from meetings to documents and board-member interactions and requiring public agencies to turn requested documents over in a "reasonable time." Sue wanted a look at the WRAP draft report. That's the report that District staff had been working on for years—the one that would support their recommendations for change to permitted water withdrawals and revise the water-supply system's management.

Sue was one of those seasoned reporters who appreciated a good tip. Sue didn't come up with the idea of a public-records request alone. She was too new to the issue. Someone close to the water issue probably suggested this story, someone who didn't want to ask for the report directly. That's how this war of words and tactics was played—through media and intermediaries.

***

I'd been at the District for about six months, but this was the first I'd heard of the WRAP, so I called executive staff to find out what the big deal was.

Their reaction told me that my briefing would be dramatic and traumatic for everyone concerned, with the same effect or greater than Steve Monsees's presentation. They assigned professional geologist Andy Smith to provide a draft copy and review it with me. It's a good thing I had time to catch my breath before meeting with Sue. I didn't need a technical background to understand the implications of what Andy said.

Andy had bright-red hair, an occasional bright-red beard, and crystal-blue eyes. If he had a redhead's fiery temper, he kept it well hidden. When he was angry, Andy's eyes flashed, but rarely was he anything but kind and calm. He was a straight shooter.

He sauntered into my office, sat at the small table there, and began to take me through the report. The bottom line was apparent. The report showed that present wellfield pumping might have to be cut by as much as 75 percent for the environment to recover fully—if it could. Oh man. How am I going to find a context for this? How could Tampa Bay lose 75 percent of its water supply? How are we going to find replacement quantities, and how did the District overpermit so much? My job was to make sure that when Andy briefed Sue, she understood that we're a long way from turning science into policy and that the report was just a draft.

On June 15, the *St. Petersburg Times* frontpage headline above the fold read "Area's Water Bill May Be Coming Due." The subhead: "A new study says pumping of wellfields would have to be cut 75 percent to halt environmental damage." The story began: *"Pumping at more wellfields in the Tampa Bay area would have to decrease by 75 percent to halt the devastation of lakes and wetlands, according to the preliminary results of a seven-year study. But residents who have been hard hit by the effects of overpumping shouldn't look for relief any time soon. And water users don't have to worry about the tap running dry. It's a long way from the scientific study to the decisions by policy makers that will dictate the solutions."*

As a result of the article, de la Parte filed yet another lawsuit against the District, naming Executive Director Peter Hubbell and a permitting director as defendants. The suit claimed that the District withheld public records from St. Pete and Pinellas as they prepared for a hearing before the

Governing Board on June 28. They hadn't received a draft copy of the WRAP. They were incensed. In a follow-up *Tampa Tribune* article, I called the lawsuit unnecessary and wasteful. It was a theme we repeated for the next two years.

"We had a program to brief the board over six months," Hubbell said. "Some months, we were scrambling to get the next chapter written. Before we briefed them on the WRAP, they needed to understand leakance, transmissivity, and lots of other technical material." *Leakance* and *transmissivity* are terms that define different ways that a water body can lose quantity. *Leakance* suggests that there is a more open relationship with the system under the lake, allowing water to readily move through the lake bottom to the surficial aquifer or Floridan Aquifer. Transmissivity is the rate at which water moves horizontally.

Hubbell maintained that the board was pushing hard to get information, and more of the technical staff were feeling they didn't have time to fully process their data. "The board was pushing us, and we were pushing technical staff," Hubbell said. But not everyone saw it that way.

"The WRAP had been virtually complete since 1992," said District hydrogeologist John Parker. "But it was pretty dramatic. There were major out-of-whack things going on at the wellfields with sustainability. The Authority's models were grossly underestimating impacts, and we had the data to show it." He'd written to the Authority in 1992 to tell them so. By 1994, Parker and the technicians working on wellfield permitting were clear that the District needed to change the permits—to reduce permitted withdrawals. Parker wrote a letter to the Authority about the impacts on Lake Loyce and the Authority's plan for augmentation. Augmentation is a technique to recharge a water body or wetland impacted by low water levels. It requires a permit.

The District frequently insisted that this should only be a short-term response. Some say it amounts to robbing Peter to pay Paul—tapping the aquifer to put water on the surface when aquifer drawdown is what has already impacted the surface. Augmentation supporters say because the surface and underground are connected, it is not a problem. Even so, their chemistry is different.

District scientists saw evidence that water-quality differences between the aquifer and the surface created issues with radium bioaccumulation. The matter was under investigation. Augmentation also occurs when water from one lake is transferred to another. Again, differences in water quality can change the chemical characterization and the biota of the receiving water body. According to Parker, Executive Director Peter Hubbell wrote a follow-up letter that said essentially, "Please ignore John Parker's letter."

"I was too bold for the situation, you know," Parker said later. But he understood where Pete was coming from. "To make a difference, you have to stay in the saddle, and it's a pretty high saddle to ride."

"Look, you don't change an agency on a dime," Hubbell said, irritated at the suggestion that things didn't change fast enough. "I know there were staff who thought we ought to go a lot quicker, but I was reading the political tea leaves, the board politics, and you don't turn this around in a week." Hubbell was angered by the suggestion that the District or he ducked controversy. Both agency and man took a lot of political heat all over the state and from many sectors for the changes they were trying to make.

And yet they pressed ahead. St. Petersburg didn't want to be compared to Pinellas and the Authority, so I regularly paired them in public comments. It was a strategic choice to make them uncomfortable with their litigation partners, to get St. Pete to differentiate itself, hopefully driving a wedge between them. I've always suspected it was Ed de la Parte that suggested the WRAP story to the *St. Pete Times*. I did my own story planting, and, later, I did some things that I promised myself I'd never do again.

Chapter 9
The Emergency Order

Betty Tillis wasn't homeless, but the place she lived was hardly a home. Located in the Shady Hills subdivision of Pasco County, Florida, her residence was nestled among tall trees, far from the noise and busy streets of a city. It was her choice. In 1993, the well providing water to Betty's home went dry. Betty maintained that the thirty million gallons per day of water withdrawn from the nearby Cross Bar wellfield was responsible for a drop in the water table—the drop that caused her well to pump sand and fail, as the technicians say.

For Betty, the $7,000 it took to drill a new well deep enough for water was too much. She and her husband couldn't afford it. If it weren't for the hose running from her closest neighbor, her mother-in-law, Betty would have been totally without water—not homeless, but not able to call the place she lived home. From 1993 until 1997 or so, she cooked, cleaned, washed dishes, showered, and flushed the toilet—with water from a hose.

If she could prove that the Cross Bar wellfield caused her private well to fail, Pinellas County or the Authority would have to buy her a new well through its Good Neighbor policy. But Betty couldn't prove it. And experts disagreed. So Betty and her husband were without their own source of water for years.

The Good Neighbor policy provided that within a certain geographic distance around a wellfield, private wells would be replaced if they went dry or if they failed. The policy got mixed reviews from both sides of the issue. First, the geographic area, sometimes known as the "magic circle," was arbitrary. Second, local governments sometimes resisted replacing wells because government wells were in place before the private wells, and the private wells should have been dug deeper. "First user" is part of Florida water law, where existing legal users take precedence.

"It's not so bad," Betty once said with just a twinge of bitterness in her tone. "When the hose is over here . . . we know when they wash clothes, they know when we go to the bathroom. Of course, we can't shower at the same time . . . but we've worked it out. It could be worse."

It was hard to imagine how.

"We tried to negotiate through the issues," District executive director Hubbell said, remembering many long meetings and late nights. "But we couldn't figure it out. If we can put a man on the moon, why can't we come to terms with this? This isn't rocket science."

Still, no resolution was forthcoming. While St. Petersburg's representatives were angry, Pinellas and the Authority adopted a scorched-earth program, according to Hubbell. "It was difficult to get any movement." He said he was naive, believing that if the District figured out the science and provided good technical backup for it, the local governments would work with the District to fix the problem.

Unfortunately, the local governments that got the water didn't believe there was a problem. "A lot of times the only recourse was to say, you know, that's crap, and here's why," Hubbell said. "We reached a stage where compromise didn't work, because if it were left to Pick [Talley], Eddie [de la Parte], and Chuck [Rainey], we'd still be litigating today. How do we go up against a powerful force, a set block, and change it? We needed to take our data to the court of public opinion. So we did."

As it became abundantly clear that there would be no negotiated resolution, District lawyers looked for alternatives that wouldn't tie them up in court for years. They wanted to do something that had teeth right now, something that would take the delay tactic away from the other side.

District staff lawyer Jim Robinson found it. He had come to the District from a private practice in Maine. He liked water law, and among the District lawyers, Jim was the fighter. He loved the challenge, and he wanted to win. He found an alternative in Florida law.

"We could issue an emergency order," he suggested. If that happened, the other side would have to comply or begin real negotiations. The difference between an emergency order and a water-shortage order is that a court must ratify a challenged water-shortage order and there were two already in court. An emergency order takes effect immediately and stays in effect until a court overturns it. The emergency order is among the most powerful of the District's tools to protect water and related resources. It had never issued one before. No water-management district had ever issued one as a litigation tactic.

In correspondence faxed between the public-communications and legal departments, the District was preparing to explain why an emergency order was necessary. It was dated June 21, 1994:

> The draft emergency order for a reduction in pumpage by the West Coast Regional Water Supply Authority is a step toward stabilizing the downward trends in groundwater. The order is meant to be an interim measure as the District and West Coast [Authority] work toward long-term solutions. The order proposes a restricted level of 16 million gallons per day (mgd) which is approximately mid-point between what the District considers "base level" and recent daily average pumping of 20 mgd. The proposed reduction is subject to changes based on discussions with the West Coast Regional Water Supply Authority and its member governments, and the scrutiny of the Governing Board.

Executives planned to discuss it at the June board meeting.

The draft provides the District's reasoning, its goals, and its willingness to help pay for new water supplies with a gradual reduction in reliance on groundwater sources. It was preparing to go to the wall.

The District tried to prepare the public and the media for what was to come. Both the *St. Petersburg Times* and the *Tampa Tribune* carried stories quoting Pinellas County utility director Pick Talley threatening building moratoriums if the emergency order was adopted. Pick Talley, speaking on behalf of Pinellas County economic interests, assured people that a 4 percent cutback—proposed by District staff—would severely damage the local economy.

"Besides," he said, "the cutback would create only the tiniest benefit to lakes and wetlands around the wellfields because the five-year drought—not pumping—caused most of the damage."

The District responded that cutbacks were the only way to combat a pattern of environmental damage spreading out from the wellfields. With a public report indicating that a 75 percent cutback might be necessary, a 4 percent reduction represented something doable, a good first step.

But Pinellas (speaking through Talley) wanted none of it. The District labeled the Pinellas building moratorium inflammatory and observed that the requested cutback amounted to only four gallons of water per person per day—one flush of the toilet. It seemed like such a small request—and yet, Pinellas soundly rejected it.

***

One morning I was scheduled to do the popular WFLA morning radio program *The Jack and Ted Show*. Longtime residents of the Bay area, hosts Jack and Ted were aware of the periodic battles over water, as was their producer, Sue Treaccasse. It was clear that the water wars were heading toward something big. The conflict was in the news every day. The rain hadn't yet started, and the environmental consequences were evident to everyone. My job was to describe how bad things were for the people who lived near the water-supply sources. We were asking for voluntary public support for conservation. We asked people to reduce use by four gallons per day, or one flush of the toilet. What could be more reasonable? It wasn't yet 7 a.m. when the phone rang at my house. Sue called to put me on hold so I'd be there when Jack and Ted were ready to take the call.

"Hey," I said, answering the phone, still not quite awake. I am not a morning person.

"Hey," Sue responded. "Jack and Ted will be ready in about ten minutes. Can you hold?"

"Sure."

"You'll be on with Commissioner Rainey," she said, and she was gone.

My heart skipped. Commissioner Rainey? I tried to be calm but my thoughts were racing. Staff at my level were never to oppose an elected official (it was a new rule I had made up). Rainey was the most powerful man in water policy. I was District staff. Rumors of his demands for staff departures and reprimands were part of District culture. District deputy executive director Dave Moore smiled proudly when he told me how Rainey demanded his firing and how he survived. Could I survive a threat from Rainey? What could I do? I could reasonably hang up and say I was disconnected, but then Pinellas would have the morning show to itself. I could use the fax line to call an executive to do the interview, but it was my job, and what if I couldn't find one in time? I got a grip on myself and thought, *"You can do this; you are no wimp. You are doing the right thing." And a final self-admonishment to myself: "Be nice."*

Sue popped onto the line and into my thoughts.

"Thirty seconds, Honey."

Jack and Ted came to me first. "What's the situation out there, Honey?" I gave a quick overview of the environmental problems, then reiterated that the District didn't want to litigate. The District wanted to develop new water supplies. "Present sources are being used beyond capacity, and people and the environment are suffering for it," I told them and the audience.

"What do you say, Commissioner?" Ted asked.

"Swiftmud can't prove anything," he said. "They never have been able to . . ."

Before he could finish, I jumped in. I was angry.

"Commissioner, we have boxes and boxes of proof. We have rooms full of boxes of proof. But you know what? That means nothing to the mother who lives in a trailer next to one of those wellfields, who woke up this morning to find her well dry, and who can't mix formula for her baby."

There was silence. My thoughts caught up with my mouth. Uh-oh.

I don't remember what Jack or Ted said, but the conversation petered out and then we were disconnected.

I was giddy in the way you are when you don't yet know the consequences of your actions. I liked the job at the District. I would miss it, I thought.

Getting dressed, I tried to think of my game plan for the day. In the car on the way in, I got a call from Assistant Executive Director Mark Farrell. Thankfully, he was howling. "I almost drove off the road," he told me, barely able to catch his breath while laughing. "It was beautiful, Honey. A mom, formula, a baby. Dry wells. You had him. He came off as Attila the Hun." Mark was exhilarated.

But Mark didn't know what I did.

It was a true story.

***

When staff for the Authority and the District met on June 22, 1994, *St. Petersburg Times* reporter Sue Landry was there.

"Accord in Short Supply as Water Agencies Meet," screamed the bold headline the next day. The subhead: "Regulators won't budge on plans for an emergency cutback; the supplier won't budge on dire predictions if the plan goes through." It was an intense game of brinkmanship, with each side determined to make the other back down and everyone assuming someone would; but no one did. "The two sides disagree on nearly everything except that lakes near wellfields are drying up in the midst of a five-year drought," she wrote.

Sue neatly drew a clear distinction in the values of the two sides, capturing the situation well.

District assistant general counsel Rick Tschantz said, "We're talking about dead species, animals, fish, and aquatic life." To which Pinellas utility director Pick Talley responded, "I'm talking about lives, jobs, and businesses. I think we can sacrifice a few fish for that." What he meant, of course, was that Pinellas County was willing to sacrifice a few fish in Pasco County so that lives, businesses, and jobs in Pinellas County were unaffected. But it was more than fish; it was property values and public health and safety. The remarks were cold, but it was classic Pinellas, an echo of the 1970s.

Still trying to force the District into an untenable and unpopular position, Talley suggested that if an emergency existed, why not order a ban on outdoor watering? The District kept insisting that if we'd quit arguing and treat this as if a water main had broken, the public would understand and support additional conservation measures.

District assistant executive director Rich McLean summed up the meeting. "We're talking about change," he said. "We're saying you can't keep doing things the way they have been done." For another week, headlines raged about home values, the cost of a moratorium, and the economic consequences of an order to reduce pumping.

Threatening building moratoriums and trying to scare the powerful development community into acting, Pinellas County (through lawyer Ed de la Parte and Utility Director Pick Talley) tried to move their stories forward. *Times* writer Sue Landry quoted Rich McLean saying that a building moratorium was technically possible but that it was unlikely.

***

To clarify his statements, I asked Rich to put his name on an op-ed responding to the building-moratorium argument. He didn't hesitate.

Crammed into the public-communications office at the Tampa Service Center, Assistant Executive Director Gene Schiller, Executive Director Peter Hubbell, Assistant Executive Director Mark Farrell, and I sat and talked about what we needed to do. The office was meant for two people, three if you were good friends. The four of us were practically on top of each other. I sat at the computer; the executives sat at the desks were jammed head-to-head. The building-moratorium argument wasn't gaining traction, but we felt compelled to respond, particularly after Rich's comments to Sue Landry. It was a delicate balance.

We couldn't deny he'd said it—he had. And I didn't want to make Sue mad by suggesting she had done something wrong. She hadn't. So we needed to expand our response. We'd already publicly shamed Pinellas for its scare tactics. Assistant Executive Director Eugene Schiller brought up the issue of the "big lie"—like Hitler's Germany. He said, "If you're going to tell a lie, tell a big lie." We thought Pinellas was telling big lies. That's what we titled the op-ed, and it was our lead. We didn't hold back. We blasted them. I drafted it with input from technical and executive staff, fact-checked it with Rich, and tone-checked it with executives, and out it went. The points were strong: *"It's the big lie. Scare tactics. By frightening people with the outrageous claim that a building moratorium is the only way to save enough water, some utility directors are playing a dangerous game that does not serve the public interest. The fact is that any person can see that saving three to five gallons per day per person is not too much to ask and that most people who care enough to save four gallons, will end up saving much more. We can protect the environment, the economy, and our quality of life."*

***

The barrage of op-eds put out by the District and the strong language in them provided a clear indication that things were different.

From a public-communication perspective, the District held nothing back. From a legal standpoint, the Authority, Pinellas, and St. Petersburg held nothing back. Each side, in its own way, decided this war was not going to be resolved, and each pressed its case with what it believed was a winning strategy. The District had to fight for the emergency order before it got to the Governing Board. Pinellas and West Coast pulled out all the stops, going to court to stop District staff from presenting the emergency order to the board. Everyone felt that this was do or die—including District staff. Still, the political pressure and the distractions were tremendous. In 1994 alone, the District underwent two audits, and two investigative TV reporters from two channels ran multiday features—one of whom continued his reports for eighteen months. Legislative review came from the Water Management District Review Commission, a federal investigation looking at the Sarasota National Estuary Program, and several investigative newspaper reports. It was hard to imagine these were coincidences.

The only thing that Pinellas's representatives admitted is that they charged the Governing Board with violating the Sunshine Laws. During a deposition, a District employee suggested that improper communication occurred among two or three board members.

Sunshine Laws prohibit elected and appointed officials from discussing or conducting government business when and where the public cannot participate. Board members may not call or correspond with each other or use a staff member to carry messages from one to another. The District and at least one Governing Board member thought it was a smear campaign. It is little wonder that when the District got the opportunity in 1996, it responded in kind. All these moves were meant to distract and discredit the District, Assistant Executive Director Mark Farrell thought. They were also meant to discredit District leadership, but it didn't work.

"Some of those people were trying to bring down the water-management district altogether," says Farrell. "They came in and demanded staff be fired, they went to the Florida Senate to have Governing Board appointments denied, they had deep pockets, and they went after us from all sides, but the staff didn't cave. And neither did the board."

Few people noticed the small item that appeared in the *St. Petersburg Times* the morning of June 30, 1994—the day the Governing Board was scheduled to consider the emergency order. Without a byline, the brief said, "Pinellas County Commission Chairman Bruce Tyndall is trying to organize a water coordinating council which he hopes could act as a peacemaker in a growing dispute over water. Tyndall said he was looking for discussion, unencumbered by statutory responsibilities."

It was an idea whose time would come—and when it did, the District welcomed it.

There was standing room only in the District boardroom on June 29, 1994. In June, Florida is hot even early in the day. Still, all the lawyers, engineers, hydrologists, and spectators were dressed in suits and ties. The reporters and the activists wore jeans. The building buzzed with expectation. Something big was going to happen that day—either the District would act, or it would back down. Either way, everyone enjoys a public spectacle, and this promised to be a good one.

By business standards, the District boardroom is not fancy. The light oak paneling, blue padded chairs, and recording equipment were good, but not top-of-the-line. Occasionally, on a slow news day, the District answered media questions about the cost of each item, and much would be made of this Taj Mahal, as it was dubbed. The $1,400 price tag for board chairs seemed like a lot, but anyone who had to sit there for hours on end needed a comfortable—and durable—chair. The building would not be noticed in Tampa, but in the small city of Brooksville, it was.

An hour before the meeting, an appeals court ruled that District staff could present the emergency-order option to the Governing Board. It was a first, definitive loss for the Authority, St. Petersburg, and Pinellas. District staff were elated. Staff thought they had the votes to get the emergency order passed, but it was anyone's guess how the day would end.

Rick Tschantz, District assistant general counsel, opened his presentation on the emergency order, giving a brief description of the morning's legal decisions. He was several minutes into the presentation when Pinellas's lawyer, Ed de la Parte, stepped to the microphone to interrupt. De la Parte wanted more time to present his case. He asked for an hour. The board discussed it. Chairman Charlie Black offered thirty minutes. Vice-Chair Roy Harrell argued for the hour. They all agreed that it was important, but they wanted to ensure that anyone who cared to comment could. The room was full and spilled over to a second room; the blue cards were stacked high. When the public want to address the board, they fill out a blue card. The cards entitle speakers to three uninterrupted minutes. John Hamner, the board member from Manatee County and a former newspaper editor, said, "I have no desire to sit here all day looking at maps and charts and graphs. As I understand it, we're giving staff permission to move this to a court."

"Mr. Chairman," de la Parte said, with deference, "what you're doing here today is much more than deciding whether to send this to an administrative hearing. It will have a profound effect on the one million people of the Tampa Bay area who will be affected before we have our administrative hearing. What you're doing here today is denying us our administrative due process rights. For those reasons, we ask that you please take the time necessary to hear us out and hear the facts before you take action on the staff recommendation."

His request was denied, and he was asked to refrain from further interruptions. Tschantz went on with his presentation. He reminded the Governing Board that the Authority and Pinellas County had filed motions to disqualify the Governing Board from hearing the emergency order. As was customary for de la Parte's firm, motions were filed just at or just after 5 p.m.

District general counsel Ed Helvenston made a lengthy presentation on the different motions and filings of Pinellas and the Authority. In the end, the board moved to dismiss or refer the motions to a hearing officer. It was ready to hear the emergency order.

Rich McLean, deputy executive for resource regulation, briefed the board on the technical issues, including the environmental reasons that created the emergency. He deferred to other experts who detailed the impacts, the hydrology, and irreparable harm to fish and wildlife.

"All the species that can get out do," McLean said. "The rest die. What we do is protect those things that cannot get out." He and the others showed lots of pictures, lots of videos. They wanted to show and tell like Steve Monsees, whose hundred-acre lake had gone dry. They wanted the board to see what they were talking about as much as to understand it. Florida Statute 373.119.2 states: *"Whenever the executive director with the concurrence and advice of the Governing Board finds that an emergency exists that requires actions to protect the public health, safety or welfare, the health of animals, fish or aquatic life, a public water supply or recreational, commercial, industrial or agricultural, or other reasonable uses, the executive director may issue an order, without prior notice issue an order reciting the existence of the emergency and requiring such action be taken as the executive director deems necessary to deal with the emergency."*

District lawyers wanted the Governing Board to know that it had the authority to take this extraordinary step. This unprecedented step.

"The system could crash because of the violation of regulation levels," District attorney Jim Robinson told the board members. He was talking about the environmental system. He said that the emergency order was required based on the protection of the public water supply and the health of animals, fish, and aquatic life. He referred them to the executive order in their board packets. As they rifled through their papers, he pointed out that the District was asking that the order take effect immediately and that, according to statute, the order be immediately binding.

It was the District's ultimate trump card.

"It is the staff's recommendation," he said firmly, "that the Governing Board advise in favor of the entering of this order by Mr. Hubbell and that they concur in the entry by him."

Chairman Charlie Black asked for questions. There were none. A break for lunch was called with de la Parte scheduled to present afterward. St. Petersburg's attorney tried to defer part of his thirty minutes to de la Parte but was informed by the chair that "he's not going to get thirty minutes . . . he's only going to get ten." There was the slightest bit of annoyance in the chair's tone.

It was a very short break.

"I want to discuss the fairness and equity of the order proposed," de la Parte began. His sonorous voice was smooth, his discussion clinical. He spoke clearly, but his arguments were complex. Worse, it was hard for the Authority, Pinellas, and St. Petersburg to argue for fairness when they had adopted a scorched-earth posture of aggressive defense and denial.

"It targets public supply in the central system," he said, "leaving other users out of the equation." He provided a copy of the draft WRAP and pointed out that the central system wellfield accounted for only 39 percent of the water use in the region.

"We have a photograph of Starkey Wellfield, and it's just as bad," he said, noting one photograph with treefall, another badly burned. De la Parte asked, "Why wasn't Starkey included in the order? Why can't we spread the cutbacks to all users?" Mining and agriculture used more water than any single wellfield. And recreational use of the resource amounted to six million gallons.

"Your staff tells us our cutback is about six million gallons. Why not take those six million gallons from recreational users?" de la Parte asked the board, the staff, and the audience. "How can recreational use of water be afforded a higher priority in this situation? If uses are to be cut to address an emergency, then uses of a lesser priority should be at the top of your list."

The pace of his speech increased as he warmed to the subject. "Instead, it's unfathomable to me that you have put public use at the top of your list for cutbacks from this area."

He went on to address alternatives to the emergency order. He began to ask about the wisdom of issuing new permits while proposing the cutbacks. Why hasn't the District investigated and shut down all illegal uses? Why weren't all users cut back equally? Why not ban the outdoor use of water or reduce outdoor use to one day per week? These were reasonable points. Perhaps under other circumstances, if the denials of negative impacts had not been so aggressive or de la Parte was not the one delivering the argument, they might have made a difference. Under these circumstances, the arguments came across more like "Take anyone's water but ours."

After de la Parte, Pinellas utility director Pick Talley spoke. He got five minutes. When quotes are attributed to him in the paper, they seem so harsh, so stark, and so cold. But at the microphone, addressing the Governing Board, he spoke with an urban southern accent coloring his words. He was calm, reasonable, and softspoken. "I took some pictures myself that I wanted to share with you," he said. "My point is that this is a debatable issue and that it should receive a public hearing and that this process eliminates that."

He pointed to a photo and said, "This is Monsees Lake. And, by the way, a permit to augment this lake has been here for several weeks and hasn't been acted on. The next lake is unnamed. It's about eight miles away from Cross Bar. A Pinellas County employee took it," Talley explained. "He tells me that these white things are dead fish. Here's a picture fifteen miles from Cross Bar. You can see that these pictures show the same kind of stress in the pictures the staff showed you earlier."

It was very close to the argument used by Deputy Dave Moore in the staff debate. The simplified argument was that because other lakes were affected by drought, the damage to lakes close to the wellfields must be attributed to drought as well. Talley continued, showing several pictures and discussing the relative merits of augmentation, and concluded that "the issue is debatable." He gripped both sides of the podium, leaned into the microphone, and told the board that District staff had waffled between blaming drought and blaming pumping and couldn't come to grips themselves. "It's debatable," he concluded. He repeated it for the next two years.

The bell rang, indicating that Pick's time had expired. He asked for three or four more minutes. Since the chair granted his request, lawyer Kent Zaiser, representing St. Petersburg, came to the microphone to argue the point further. When he concluded, he asked the chair to allow Pick some of St.

Petersburg's time. The chairman noted, "Mr. Zaiser, he said he only wanted three to four minutes. You've already taken the time I was going to give him." Everyone laughed. It loosened up the room, but not much, and only for a moment.

Pick's argument can be summed up in this way: The District was not taking away 6 million gallons per day—the difference between present use and the proposed cutback level. The District was taking much more. Since the permit allowed up to 145 million gallons per day, it was a loss of 29 million gallons. His arguments were abstract. Even if people understood that the District was taking something that wasn't being used, it's hard to say whether they would have supported the notion. Most people didn't have the background to understand.

Furthermore, they didn't want to listen to an unsympathetic speaker. The Governing Board had the basic knowledge and had to listen, and so did the people in the audience that day. Even so, the difference between what was allowed by permit and what was now used was a chimera. If present pumping levels were a problem, it made sense that increased pumping would make it worse and should never be available.

Florida required concurrence for growth to occur. That is, the necessary infrastructure (roads, schools, water, and other public services) must be in place. That rarely worked out, and it upset the people who lived in high-growth areas. In the Tampa Bay region, water-supply development never kept pace with rampant growth and the water demand that came with it, except on paper, in plans that were never implemented. Because the District issued permits at quantities they later learned were too high, the planned water supply, when cut, would not meet the planned demand. In actuality, pumping levels to meet existing demand were too high, cutbacks in permitted quantities were a paper problem, but reductions in actual pumping were a real water-supply problem.

Opponents were talking about legalities, while the District was talking about dead wildlife. To the public, there was no contest. Kent Zaiser spoke for St. Petersburg. He read from prepared remarks in a monotone. In contrast to Pick Talley and Ed de la Parte, he seemed less committed. Zaiser was a former general counsel for the District. But this board was not his board. His board was gone. These people were mostly new. They didn't know him, and he didn't know them.

In his presentation, he said, "St. Petersburg supports and adopts the remarks of the West Coast and Pinellas representatives." Later, he said the city wanted to cooperate, didn't want to litigate, and generally believed that District staff had been less than forthcoming in their handling of the St. Petersburg permits. The message was confusing. The city claimed to want cooperation, yet it had actively participated in every challenge to the District's proposals.

"The environmental problems were discussed as early as fall of 1993," he said, "perhaps earlier." The city was willing to deal with those issues in the permitting process. No emergency existed. Or words to that effect. He observed that the District was already under review from the statewide Water Management District Review Commission. The commission had been established to look at the water-management districts and their operations throughout the state. It created a public platform for supporters and opponents alike.

Perhaps Zaiser was making a veiled threat to the District. His rapid, clipped monotone and rote presentation style didn't allow much room for interpretation or empathy. It was harder to follow as he droned on. The messenger botched the message.

After Zaiser finished, public comment opened. Mary Coleman read a poem that had been written by several people:

> You. . . Me. . . We. . .
>
> Who to blame for the present water crisis?
>
> Who to blame for dead trees and dying grasses?
>
> Who to blame for drying lakes and ecosystems?
>
> And. . . for the draining of our aquifer?

"Not I," said the weatherman, "the rain is too far west."

"Not I," said the West Coast Authority, "we're trying our political best."

"Not I," said the Swiftmud staff. "We're in a phase of drought."

Not I. . . Not she. . . Not he. . . Not them. . . Not we.

Who to blame for the pollution in the bay?

Who to blame for wildlife going away?

Who to blame that the green industry's folding?

And. . . for the small man, tired of holding?

Who to blame power industries waste so much?

Who to blame large companies go untouched?

Who to blame that we didn't act more swiftly?

And. . . why isn't there anyone listening?

"Not us," said the developers, "we followed building code."

"Not us," said the councilmen, "we had to build that road."

"Not us," said the county commissioners, "we just followed others' rules."

No, us. . . Not she. . . Not he. . . Not them. . . Not we. . .

Who to blame for the shortsighted planning?

Who to blame for the litigation and infighting?

Who to blame for man's misunderstanding?

And. . . for the media not educating?

Who to blame for the grower who gave up trying?

Who to blame for the lake owners all a-crying?

Who to blame for water running down the street?

And. . . landscapes dying in the heat?

"Not us," say the people; "it's the bureaucrat's job."

"Not us," say the businessman; "it's me they want to rob."

"Not us," say the engineers; "the blueprints are all here."

"Not us," say the politicians; "this isn't an election year."

"It is us," said my generation. "We slept on our watch."

Not she. . . Not he. . . Not them. . . But you. . . yes, and me. . .

We all are to blame, for the unfulfilled tomorrow.

And we all are to blame for creating this dilemma, this sorrow.

We all are to blame. . . We all are to blame.

You. Me. We.

A round of applause punctuated the conclusion of her remarks.

Speaking for the Coalition of Lakes Association (COLA), a lawyer said, "West Coast and its member governments, in response to the District's determination of environmental damage, refused to undertake the examination of alternative supplies and instead pursued extraordinary legal actions to insulate themselves from taking responsibility. We think it's time for the board to stop taking direction from West Coast Authority, stop allowing delay, and while it's probably not enough, take action today."

Next, a distinguished-looking man who worked for the Pasco County School Board rose to speak. Bill Alexander's anger at the government and his self-imposed restraint warred inside him. Occasionally, his anger won.

Barely concealing his outrage, he told the Board:

> I think I'm the first person to address you today who's not being paid to be here. I've exhausted most of my vacation time for this year, to attend government meetings on this subject. If my words today are critical and angry, it's because while we applaud your efforts today, it's too little, too late. We are rational. We do believe in the system. There are people who have worked hard their whole lives to get a pinch of paradise. They are not going to sit idly by. Why are we angry? We pay taxes to pay West Coast, we pay taxes to support the water-management district, we pay for the staff who've made the presentations, and we must buy fish to put in the lakes to eat the weeds, we must organize associations to fight, and we must hire attorneys to argue our case . . . and at the same time, I have to buy flood insurance. Do you understand why some people are angry?
>
> We have sins of commission and sins of omission. We've had lots of both. You guys must have the wisdom and the courage to make the difficult decisions. We would have never let this happen, but it happened so slowly, insidiously under our noses. But it didn't happen overnight. We don't need a scientist to tell us that. I ask you to be strong, to be courageous. We ask you to kick your staff and tell your staff not to be afraid and do the things necessary to preserve our quality of life. I find it interesting to hear the speakers today talk about the fish as if they are different. They must have gone to college before ecology courses were offered. They do not understand the interdependence of life. Well, we do. We hope we have your support.

Representing the green industry, Earl Wells said, among other things, "It's time for all segments of the community to share in the misery. Conservation across the board," he argued, "will make the difference. People are part of the problem." Green-industry leaders had figured out that the District was not the enemy—that Pinellas and the Authority had used them. They came out in force to address the board and to speak in support of alternative supplies. And they came to encourage the District board to leave watering restrictions alone.

Pauline Ellis is a statuesque woman. Silver-haired with piercing crystal-blue eyes, she has the kind of demeanor that demands she be taken seriously. She was Steve Monsees's neighbor. She'd lived there since 1987—several years before Steve arrived. She came to tell the board about what she had seen:

> I live on Lake Loyce. The loss of that beautiful lake along the wetlands and wildlife has been devastating to us. The reason for the loss—the overpumping from Cross Bar—doesn't make it any easier. But as they say, nature abhors a vacuum, and with the loss of the normal wildlife, we have had several new species appear on the scene. Among those we've seen is the West Coast ostrich, typically seen with its head in the sand only to raise it occasionally to say, "It's the drought, it's the drought . . . we're not responsible, there's nothing we can do about it." Then there's the Pick Talley, more commonly referred to as chicken little, which runs about forecasting complete economic disaster and disease if people are required to cut back by even one toilet flush. And for some time, we have been following the double-jointed Swiftmud hen. This confused fowl tries to bend both ways and consequently doesn't get anywhere . . . it won't last if it doesn't stiffen up its joints.
>
> It's too bad we can't get these birds to work together in our area . . . but they seemed to be so concerned with territoriality and one-upmanship, especially the ostrich, that the concerns of the environment and wildlife get lost in the shuffle. After all, what are a few fish compared to economic growth . . . the litigation and the wild charges in the paper are certainly not the solutions but are only adding to the problem. We plead with you as conscientious, responsible officials to enact this emergency order.

The day dragged on, and the testimony continued, twice over, from every perspective. Some said the order didn't go far enough; some said it went too far. Some said the District was too late, but they said to hurry up. Another long meeting, another tough decision. Roy Harrell was the last to speak. He sounded sad and tired. He tried to find some way to explain how the sides had gotten so far apart, and he concluded, "I wish we didn't have to pass this order, but I don't see the alternative . . . I move adoption of the staff recommendation." Seconds ticked off, and the chair called the question. All in favor, say aye. There was a chorus of ayes. All opposed?

Silence. These are hard decisions with potentially millions of dollars in consequences. Each board member, each staff member, and everyone involved had to gut-check their choices. In some cases, the personalities and their histories drove choices, but for most these were decisions of conscience driven by their personal values and their sense of trying to do the right thing for the most people and for the environment based on the mission.

Chapter 10
Getting There

The night the emergency order passed, District lead litigator Jim Robinson picked up the ringing phone in his office. A familiar sound filled the earpiece. Whoooooosh. It was the sound of a flushing toilet.

***

"I thought we'd get there," Assistant Executive Director Mark Farrell said about passing the emergency. "We had so much momentum . . . I mean we had to be careful, too, but we were on this incredible roll." Mark pointed out that every time St. Petersburg or Pinellas tried to oppose District action, it backfired. There were negative headlines, letters to the editor, editorials, and more. Public support was clearly with the District and clearly against—something.

Environmental damage? Failed water policy? The lack of compassion for affected residents revealed in public remarks? Whatever the reason, for the first time, the District was enjoying broad support from the public, including most residents of Pinellas County. The staff were elated. There was a sea change in morale. The staff felt they were doing the right thing. They were serving the public interest. Of course, not everyone saw it that way. Some suggested that the District focused on Pinellas, St. Petersburg, and the Authority to the exclusion of everything else. They hinted that the District might win the battle but lose the war.

"The District had pretty much sold the public that the wells were creating environmental damage," St. Petersburg mayor Dave Fischer said. "But my team was telling me that it varies depending on the wellfield and that you can't just make blanket statements about the whole thing. So I went to look at the wellfields, and I saw some of the damage, and the experts said it was uneven. . ." his voice trailing off as he thought about what he saw at the wellfields.

The complicated technical and political issues captured the attention of elected officials and experts around Tampa Bay. Neither side had a clear idea of where it would all lead, and both sides knew the consequences were potentially huge, economically and personally.

"I think the sum total of it all was how we could position, how we could move through logical steps," said Executive Director Hubbell, explaining the need to attempt the water-shortage orders before the emergency order. "We knew that there were actions we could take on water-use permits, but there were a couple of safety issues that were the bottom line. We could have gone either way: forcing them below permitted levels or letting them break them if we believed they were really trying. We had lost the ability to just slough it off . . . if we wanted to turn our backs. We had cause-and-effect data."

Of course, no one wanted to turn their back at that point. District staff were focused on getting the systems—water supply and environmental—back into balance.

"We had a very focused goal. We had everyone in the agency pointed in one direction." Hubbell added, "And that's a pretty powerful force." The system was out of balance, he said, and to fix it, "people had to look at things differently."

But Pinellas, St. Petersburg, and the Authority didn't want to look at things differently. They believed the District had abruptly changed the rules and found its actions outrageous and capricious. Even if there were problems at Cross Bar, de la Parte argued, the District had done more than issue the permit. The District had insisted that Cross Bar was an excellent site for a wellfield! In interoffice correspondence dated February 1, 1980, Loretta Hennessey, District hydrologist and permit evaluator, pointedly wrote, "It is good to see wellfields located in drier upland areas where the biological impacts are much less than in valuable wetlands like Cypress Creek." Her conclusion, written on the next page, was less than visionary: "There should be no long-term adverse impacts to the biological systems of the property adjacent to the wellfield. Impacts of pumping will be centered within the wellfield."

More than a decade later, Steve Monsees and his neighbors were living with her grievous error. The mistake was the District's, de la Parte pointed out. In good faith, Pinellas relied on that permit and the water it represented for its residents. Pinellas had authorized construction, growth, and uses for the water. If the District erred, the District needed to fix it—but not by punishing Pinellas County, the Authority, or St. Petersburg.

After the emergency order was passed, the District commissioned a study to measure what the public was thinking about the entire furor. In the summer of 1993, the District commissioned a survey that showed that it had less than 30 percent public support. In 1994, no one had any idea what to expect, but we needed to know what people were learning about water resources, if anything. In the report prepared by a survey company and drafted in July 1994, the District's support was a whopping 74 percent. Unbelievable. More important, for the first time, people understood that the water shortage was real and that conservation is not only for some future value but for a very present need. We were very excited with the results. Clearly, we were meeting public expectations, and clearly, our pleas for support were being heard. We did what Hubbell said needed to be done in his comments following Steve Monsees's presentation to the board: "We need to link conservation to environmental damage."

In the days following the emergency order and the extensive news coverage, a *St. Petersburg Times* headline told the story: "They Get It." Pumping had dropped from a high of 125 mgd to 120 mgd in less than one week. Just as the District had said all along, the public was capable of understanding the problem and helping—when there's a clear message.

Litigation as Communication

"The emergency order let us do something immediate and definitive," Mark Farrell said. "They couldn't block us, and that was a big difference. We were able to send a message, and our power was exerted, and Pinellas County could do nothing but bitch and growl. Eddie de la Parte didn't have a blocking maneuver, didn't have a delaying tactic."

That was especially sweet to the District.

Pinellas, St. Petersburg, and the Authority were angered by the District's actions and by what they perceived as ongoing media support for the District's unreasonable positions. In a July 10, 1994, op-ed, Ed de la Parte took exception to a previous *St. Petersburg Times* editorial. De la Parte's explanation was titled: *"Times Playing Cheerleader to Swiftmud's Power Grab."* Well into the second paragraph, he wrote: *"Swiftmud, the District, is attempting to parlay the current drought into increased control over the water supply system serving 1.2 million people in our area. Swiftmud is also preparing the public for higher water rates, lower growth, and water rationing through a slick public relations campaign."*

In fact, the District was exerting its authority to move water policy in a new direction, one that was not supported by Pinellas County, St. Petersburg, or the Authority. The rest of de la Parte's column focused on the same economic concerns and complaints about cutbacks that Pinellas, St. Petersburg, and the Authority had been using all along—and had used in the 1970s. This conflict was about power—political power, economic power, democratic power, and the power of communication. Emotions were as high as District staff and activists' spirits.

"For the longest time it was two steps forward, twenty steps back, three steps forward, eighteen back, depending on the day," said Eileen Hart, who was among the longest tenured and most even tempered of the water activists. "Then, all of a sudden, it was like three steps forward and five steps back and eight steps forward and six back, and we were actually gaining." She smiled, remembering how things had begun to turn. "Once we got board members thinking and looking at it. . . well, they knew they were hanging themselves politically if they wanted to be reappointed, but they did it anyway."

"Finally," District geologist John Parker said. "We're no longer the apologists for all these impacts. Finally, we're trying to do something. I was surprised, pleased."

John and Eileen weren't the only ones surprised by the board's action.

"The emergency order was like Bull Run," said Steve Seibert, a Pinellas County commissioner at the time. "Bull Run because we were blithely moving along, and we got our clocks cleaned," he said, stunned by the District's actions. "The District threw down the gauntlet. I thought to myself, 'Holy cats, this is a real issue.'"

It was just the beginning. A land-use lawyer by trade, an environmentalist by choice, and clearly one of the new "compassionate conservatives," Seibert glowed with enthusiasm as he embraced a subject. Most people will tell you that Seibert is a mediator and consensus builder by training and by personality. But early on, Seibert had adopted the strident and litigious policies of the Pinellas County Commission.

Defining the Bad Guys

As a tactical communications matter, the District assumed the St. Petersburg water team (staff and consultants, lawyers, and experts) was a group that we could work with. Its permits were the first to be considered under the District's tough new environmental stance. Yet St. Petersburg went along with Pinellas and the Authority. St. Petersburg, the District knew, liked to distinguish itself from Pinellas County.

In the 1970s, St. Petersburg (along with Pinellas) resisted regulation, but later it worked with the District until an agreement was reached. Pinellas never stopped resisting. St. Petersburg was different, and the District wanted to drive a wedge between them—knowing that the link between Pinellas and the Authority was too strong to break.

So, in every public comment, in every document, in every interview, the District, that is, I, named them all together as if they were one entity, and many times, I mentioned St. Petersburg first. Reviewing the newspaper clippings, it's easy to see now. To pressure St. Petersburg to step away from Pinellas, the District wanted the city to know that they would be painted with the same brush. St. Petersburg didn't like it.

Beyond the public drama, tensions among the parties were high privately; the pretense of civility slipped away when opposing staff members came face to face. Negotiation meetings were constant and relentless.

"To this day, Kent Zaiser and Dan Fernandez think I hosed them," said Hubbell. "And it was clear that Bill Johnson hated my guts . . . They just didn't get it. The subtleties just got by them. For all the water-shortage orders and the emergency order, we never actually modified their permits."

Zaiser and Fernandez shared responsibilities as outside counsel for St. Petersburg after Ed de la Parte moved on to represent the Authority. Each was a former District general counsel.

The emergency order applied to all the wellfields in what we now called the central system, not just St. Petersburg's. Still, the District continued to discuss St. Petersburg's permits as a separate issue. The hope was that if the District could reach an agreement with St. Petersburg, it could set an example for the other local governments. St. Petersburg wanted that too, but no agreement was reached. So the District held those permits. They were not issued. The permits had been under consideration for more than two years with each side continuing to ask the other for extensions—delaying the outcome—perhaps fearing the unknown. The last extension was approved the day before Steve Monsees addressed the board months earlier.

"We had lived up to every condition in our permits," Bill Johnson claimed. "We were not evil water pumpers. We felt that we could prove we were entitled to our permitted quantities. We'd reduced our water usage to 1970s levels. What more could we do? What more should we have to do? And here the District is planning to cut our permits, and they don't even tell us directly. We found out about the District's plan for cutbacks from the newspaper," he said, disgusted, recalling the article the *St. Petersburg Times* produced because of a public-records request.

"Our folks didn't even have a chance to look at the science. Where are we going to get water? We're not evil. We're just trying to meet the needs of our community. Not one person at Swiftmud has ever been in our shoes supplying water to people. They just didn't get it," Johnson said.

Johnson recalled that in a permit-negotiation meeting with District staff, Hubbell had suggested, "What if we tie your permit quantities to lake levels?"

"Wait a minute," Bill, incredulous, had responded. "You're telling me that if a lake is down, then our wellfields will be cut back?"

"Yes."

"I'll see you in court."

To Johnson, an engineer, the idea was stupid and impractical. To Hubbell, the hydrogeologist, it was a starting point. Hubbell remembered the meeting, too. "Every time we brought up the lake levels, Johnson would say, 'But we never violated our permits,' as if that meant that no harm was done. As water managers, we look at that and say, 'Yeah. You quit pumping just before you exceed regulated levels, but so what? Look at the environment.'"

From the District's perspective, the already-stressed wellfields were worked harder during dry times when demand was up. By tying permitted quantities to lake levels, you forced a utility to consider what's happening in the surrounding environment.

"If we cut back, where is our water going to come from?" Johnson asked repeatedly. "How can you get water to people? And what about the millions of dollars sunk into infrastructure? What do you do with the assets?"

In fact, the District had a plan for that, too. It wanted to offer St. Petersburg gradual cutbacks, stepped-down quantities, while offering financial help for replacement sources. Johnson remembers nothing of that. But he was clear on the threatened cuts. He insisted the District had no management plan for water resources. "What did they manage?" he asked.

Sitting in a negotiation meeting with Rich McLean, District deputy executive director for resource regulation, Johnson pointed out the development sprouting all around St. Pete's Section 21 wellfield. "We'd been doing the same thing for a decade," Johnson said. "The only thing that was changing out there was all the new development. They've dug all these borrow pits, and when it rains, the water goes into the pits, not to the wellfield anymore."

According to him, McLean told him that surface-water permitting handled that as if it was another entity.

"Well, don't you people talk to each other?" Johnson asked. "You're not the Southwest Florida Groundwater Permitting District. You're the water-management district."

It confirmed his feelings that the District was clueless. "How much of the problem was pumping, and how much was other factors?" Johnson asked. "The District just ignored everything but pumping."

According to the District, teasing out the percentage of factors relating to environmental damage was pointless. "Pumping was a thing that could be controlled," Hubbell said. "Other factors, like rainfall, couldn't. For too long, utility directors were allowed to just pump and pump regardless of what was happening around them. We had to change that."

With the adoption of the emergency order, negotiations over the wellfield permits intensified. "I never could get them to see that we hadn't touched their permit," Hubbell continued. "It was still in for renewal. We wanted them to understand that we were in a big crunch, things are upside down as far as the environmental impact and pumpage are concerned, and we have to, at least for the short term, until we get out of this crisis, things have to be different."

But to Hubbell's mind, Johnson just didn't get it. He couldn't see the difference between the short-term emergency order and what would ultimately be done with their permit. To Johnson's mind, Hubbell didn't get it.

"We could have been the example," Johnson said, still angry years later over perceived deception, inconsiderate treatment, and lost opportunities. "We could have shown everyone how to do it, but the District refused to consider anything other than pumping."

"I think we did consider factors other than pumping, but when you look at pumping in comparison to everything else, it was the biggest issue," Hubbell said. "And it was the thing we could control."

An Outsider Steps In

With several years of District Governing Board experience to guide him, including his tax-cutting and tax-raising initiatives, Joe Davis felt prepared to be the chairman of the District Governing Board. Joe's slow-talking, soft southern accent hid a quick mind trained for the details of law and accounting. He was liked and respected by staff and board alike. Some say that it was only a misstep by Roy Harrell that allowed Joe to become chair before him. At the time board officers were selected, it was Roy's turn to be chair. Joe had quietly let people know that he was interested in the position. Because of Florida's Sunshine Laws, direct communication was prohibited except under specific circumstances. To communicate with one another about matters on which they might vote, board members had to put it in writing and send it to everyone through the staff. This created a public record. Whether in response to Joe's interest in the chairmanship or simply to lock in his own election, in April 1994, Roy sent a letter to the board, detailing his accomplishments and outlining his qualifications. Despite Hubbell's plea to rethink it, Roy sent the letter. It was not well received, and Joe was elected. Charlie Black, Sally Thompson, Roy Harrell, and Jim Martin voted for Roy; the majority chose Joe. In the spirit of a collegial board, Roy requested a second ballot. He asked that the vote for Joe be unanimous. It was. Roy became the vice-chair.

In retrospect, it is fortuitous that Davis presided as chair through the worst of the water wars (1994–96). As a rancher and citrus grower from rural Hardee County, Joe wasn't subject to the same kinds of political pressures that Harrell would have been.

"I think coming from Pinellas County actually cost Roy that election," Davis said. "There was the sense among our board that Pinellas County ran West Coast [that is, the Authority], and there was some suspicion there. Also, there was a new-board/old-board situation. Maybe it was the new-water-sources idea, or maybe people thought that someone from outside could facilitate better. Whether he was the chairman or me, it probably wouldn't have ended up any different." But Davis offered one clear difference. "I didn't have to stand for election, and I didn't have any business interests in the area, so there wasn't a damn thing anyone could hold over my head."

Since Monsees's presentation months earlier, things had moved fast. District executives, though pleased with the board's mandate for change, felt things were on the verge of spinning out of control.

"We had been moving with deliberate speed," Hubbell said. "We were getting our technical data together, we were educating the board, working with the Authority and other local governments on the Resource Advisory Committee. With Monsees' presentation the board started to get away from us and demanded we protect the little guy from the big bad utilities. Clearly it got out of the gate."

"I didn't feel like we were picking on anybody," Davis recalled. "We didn't want to harm landscapers or tropical-fish farmers, but it seemed like we were in such a bad situation and that some people had been hurt for so long, we just had to move quickly and make the proper decisions."

Among the first to request a meeting with the new chair were the area utility directors. "They brought in big boxes of newspaper clippings," Joe said, smiling. "And they complained about our public-communication effort for two, two and a half hours." Joe had long before decided that clear communication was a counterbalance to the political power of St. Petersburg, Pinellas, and the Authority.

"Everyone was on board," Farrell said. "Even the guy sweeping the floor was delivering the message." For the first time, Farrell thought, the staff were feeling good about themselves. "They felt like they were on mission, that's what they were there for. And we started getting good feelings from the public that 'yeah, you guys are doing what you're supposed to be doing.' And people couldn't do enough to help us keep it moving."

Watching these events unfold was the only sitting member of the Authority board who had been there since the beginning, Pinellas commissioner Chuck Rainey. He was for Pinellas, only Pinellas, and he didn't like the way things looked for Pinellas.

Thinking back on it, he said, "If we had only given a little bit. . ." It's hard to tell whether the end of that thought was "we could have reached some kind of compromise" or "we could have appeared to be reasonable and stopped the negative public relations." Rainey declined to clarify. Still, something at the time escaped most people, including the cagey Rainey. The composition of the Pinellas County Commission was changing. The old ways of training a new commissioner didn't work quite as well, although some traditions persisted.

"We were always a collegial board," said newly elected commissioner Steve Seibert. "It was something Chuck Rainey taught us, a legacy, if you will, of unwritten rules. You do your homework, you don't trash another commissioner, you run the county like a business."

They took pride in holding board meetings that lasted only seventeen minutes.

"Chuck had led our water issues for a long time," Seibert said. "Why would we change what worked for us? We led with our lawyers, and we won."

The District, the Authority, and the member governments all had a way of dealing with one another that started twenty years before, Seibert said. "There just wasn't enough of a crisis to change direction."

The question was, Could this crisis be sufficient to drive change?

What Rainey and others still didn't realize was that the District wasn't going to back down this time. The world around them had fundamentally shifted. Public sentiment toward the environment had changed. Pinellas and its representatives didn't seem to notice or care. That made them the bad guys—not because the District said so, but because the public did. The District observed their real behavior, then framed its meaning in terms of the environment, the water supply, and regional cooperation.

"We're getting our fannies kicked," Rainey told his colleagues during a Pinellas Board meeting. In a February 23, 1995, *St. Petersburg Times* story, Rainey said, "They're the best propagandists since Hitler."

"He's not a team player at all," said Hillsborough County commissioner Ed Turanchik. "He's so used to fighting. He reminds me of a senior member of the politburo or something. He loved the game."

Turanchik was one of the new kind of commissioners with ties to the environmental community. His education in aquatic ecology and his training as a lawyer gave him the background to press effectively for proper and workable environmental controls. Known for his support of mass-transportation initiatives, he was dubbed Choo Choo Turanchik by a columnist for the *Tampa Tribune*. He was good-natured about it. In the following years, he needed a sense of humor.

He was chairman of the Authority from 1995 to 1998.

"It was very strange to be the chair of the [West Coast] Authority while Hillsborough County was suing them," Turanchik noted.

The northwestern region of Hillsborough County had suffered the same kind of damage as Pasco County. During the water-shortage orders and the emergency order, Hillsborough intervened for the District. Turanchik tried to stay focused on one overriding imperative: the need to identify and develop new water supplies.

"The whole situation was like a giant chess game," de la Parte said. "The District would go to water-shortage orders—well, they couldn't do that—they would go to an emergency order—well, you can't do that either. They would go to 'We're going to deny the permits,' and they can't do that either. So it became like a big chess game, and the battle escalated. The District was piling on lawsuits, and the public relations people were getting emotionally involved and bitter," he said with exasperation. "It just became more and more difficult to get back to a partnership."

"It was more like a 3-D chess game," Farrell observed. "It's not like the only thing we did was northern Tampa Bay. We had sixteen counties to manage. The water wars only involved three." He remembered how his days would go. "Get up, read the paper, listen to the radio on the way in and see what was in the news, talk to the lawyers, see what the action was, then call you into my office for a huddle."

Remembering some of those strategy sessions made him smile.

He continued. "What's Pick doing today, what do we think Eddie will file? What is Rainey doing? We had to anticipate the moves and stay one step ahead. I did enjoy it. I loved the thrill of it, and in that respect, it was fun. There was a lot of pressure and a lot of risk, but we pressed on."

Uncharacteristically, Hubbell was more direct. "It was going to the wall, going for broke. We were trying to move things off center or off the way it had been done for a lot of years. I mean, the line had been drawn."

"It was us against them," Farrell concluded.

But Commissioner Rainey and those who supported his preference for unlimited groundwater were annoyed at what the District had done. The manager of the Peace River Manasota Water Supply Authority in the southern reaches of the District, Sonny Vergara, followed developments with great interest. He had a New Water Sources Initiative grant request in for $55 million. He wanted to build additional underground storage, and the District was going to make that possible.

Vergara considered Hubbell and Farrell friends, so when he began hearing serious talk from special interests and local governments about removing the two of them, he was concerned.

"The bashing from Pinellas was continuous," Vergara recalled. "The agricultural interests hadn't quit either, and the rural communities were afraid that Tampa Bay was going to suck their water away." There was a clear intensification of the reaction to the District.

"People began to pick on the eagle with the weak wing," Vergara said. "When it could no longer fly, it was on the ground defending itself."

But Vergara hadn't seen the public-opinion results, the news reports, or the favorable editorials, and even if he had, they would have meant nothing to him. He was a political animal—one wounded by walking away from his job at the St. Johns River Water Management District years before.

The District was not a weak-winged eagle—the District was soaring with public support. As Vergara would explain his position later, "Public opinion doesn't matter." To the District staff and the Governing Board, public opinion mattered very much. Public opinion made all the difference. The District had what it wanted most of all, what it felt it needed to make necessary water-policy changes—widespread public support.

The results of the July 1994 public-opinion survey conducted in Pasco, Hillsborough, and Pinellas Counties were remarkable. More than one-third of those surveyed could explain what the District had done to protect water resources. Of those, 74 percent supported the effort. The District had moved from less than 30 percent support to 74 percent in a single year. Things were rapidly changing. Before it was over, public support for the District's initiatives went even higher.

Doing the right thing in public policy doesn't always garner public support. In this case, as the board and staff tried to do the right thing, getting the water-supply system and the environmental system in balance, the public agreed. But relying on public support when the politics of the matter are so treacherous is a dicey posture. The public can move politics, but it's a much harder undertaking because the public aren't a monolith of values like a government board or a business association. Public response to a public policy is more diffuse. Getting the public to engage with focus is a complicated and time-consuming process.

Chapter 11
Water Policy Initiatives and Interference

Gilliam Clarke was startled awake in the middle of the night. It wasn't bad dreams that brought her out of sound sleep; it was a waking nightmare. The tremors under her home and those of her neighbors were the result of pumping at the Cypress Creek wellfield, she believed. Her entire home was sinking into the ground, not dramatically, the way Florida's big sinkholes open and swallow a building, road, or sometimes a house, but slowly—each tremor an indication that the water withdrawal from under her home was causing the limestone layers there to fold in on one another, slowly lowering the entire area. Sometimes the movement wasn't slow. Sometimes the tremors were so strong they jolted Gilliam and her husband, Silbourne, out of bed at night. Like most of the people who lived with daily water pumping and the consequences of a failed water-supply policy, Gilliam coped by laughing. "If I didn't, I would cry," she said. Her house sank eighteen inches in as many months.

***

Gilliam and Silbourne Clarke bought their house in 1989, the year after Steve Monsees bought his retirement home miles away. In 1990, Gilliam and Silbourne moved in. As with Monsees, this was to be their retirement home, and they were looking forward to a quiet, country lifestyle. Gilliam became involved in water issues when the problem showed up in her backyard.

"One day in the spring of 1994, I was working in the garden, and a bee was divebombing me," Gilliam explained. "I'm deathly allergic to them, and as I tried to get away, my foot got caught between two roots—roots that weren't there the summer before. I stared at the roots and thought, hmmm, there's something weird going on here." When she really looked, she could see subsidence everywhere. Subsidence is the gradual sinking or settling of the ground—in this case, enough to expose tree roots. "There was a helluva lot of it," she remembered.

In the autumn of 1994, she called the Authority.

Gilliam, like Steve Monsees and others, called the Authority after talking to District staff.

"We have really bad subsidence," she reported. "I think it needs to be investigated." In short order, the Authority sent what Gilliam calls "a sweet young woman" named Patty to her home. "Patty walked around my yard and said, 'Gilliam, trees die.' Bless her heart, what could she do?" Gilliam recalled.

"Well, what's killing them, Patty?" Gilliam asked.

"I don't know," she said.

"Lack of water, maybe?" Gilliam offered.

"No, no, that's not it," Patty assured her.

"Then what is it?" Exasperation growing.

"Maybe the trees are growing out of the ground?" Patty offered.

"Yeah, right."

As with Steve Monsees, Gilliam's experience with the Authority was unsatisfactory.

With subsidence, the roots of a tree are exposed as the land caves in or sinks.

"I raised enough sand that they sent a consulting firm out here to check it," Gilliam said. That was 1994. "But the first tremors didn't come until the spring of 1995."

Gilliam put up four hundred notices around her neighborhood. She told people, "Something's up in Quail Hollow. Let's meet and talk about it." About seventy-five people showed up. Gilliam wasn't the only one having earth tremors strong enough to unbalance stacked boxes in the attic. Her neighbors felt them, too. A reporter came, and an article ran in the paper. Gilliam got a call from Steve Kent, owner of the Tree of Life Nursery.

"You're not going to believe me," he told her, "but the District is the best friend you've got."

"So I went to Swiftmud," she said, "and made a presentation to the Governing Board. I told them we were having tremors, and it was unsafe."

"Pete put us off," she noted, meaning Executive Director Peter Hubbell. "He could be so sweet and do so little, and that's exactly what he did. But I understand why," she added. "Roy," she said flatly as if it was the answer to a question no one asked.

Gilliam suggested that as the representative of St. Petersburg and Pinellas, Governing Board member Roy Harrell couldn't or wouldn't recognize the impacts. But she admitted that District staff were tremendously helpful to her. "The amount of information available through the District is mindboggling," she said. "And, without the staff to help, it's sometimes difficult to get at what's relevant, what matters most."

It is difficult for most people to understand the depth of water-activist commitment. The activists must learn hydrology, geology, and engineering, and political relationships must be observed and established. There are education events to organize, letters to write, disaster tours to arrange, and they must not be NIMBY (not in my backyard). Generally, people don't mind public-interest projects; they just don't want them near their homes.

Communities object to roads and schools, shopping centers, and business parks. In Tampa Bay, they object to water projects. Environmentalists are supposed to support projects with the least impact, regardless of their locations. They must be regional to be credible. If you do not think regionally, you risk the NIMBY label. Activists must recognize that the water withdrawals that gave their homes shivers in the night might put out a fire somewhere else or wash a baby. They must be bigger than the problem, and it's so hard. It's particularly hard when the water isn't used to wash babies but to wash cars or water lawns. It is not surprising that in due course, activists get fed up, frustrated, and angry. Their anger was sometimes expressed during the public-comment periods at board meetings.

Gilliam often lapsed into sarcasm.

"It's not so much that I'm sarcastic," she said, "as I try to lighten things up."

But some would argue that point. And there is always the question whether the humor or sarcasm worked the way it was intended. Even Gilliam admitted, "Sometimes I've made people angry with my brand of humor, but, oh well."

Testing the Waters

While litigation and negotiation moved along a parallel track, Bruce Tyndall, chairman of the Pinellas County Commission, had another idea: bring in some new faces.

Few had noticed the small clipping that appeared in the *St. Petersburg Times* the day the Governing Board passed the emergency order. But the District agreed to co-fund a professional facilitator and to establish a group of people, some elected and some appointed, to discuss the issues. While the agreement was made in the summer of 1994, negotiations to get the group started took some time. It was well into the fall before meetings began. The idea was to give presentations to this group and provide an open forum—a site of joint fact finding—to discuss water issues.

No activists were appointed to the panel, dubbed the Coordinating Council. Because the group had no mandate or independent authority, participants hoped that communication might be more open. Everyone agreed that nothing from the group would be used in litigation. The District was optimistic. When Mayor David Fischer of St. Petersburg agreed to participate, District staff were elated. They felt strongly that if he could get some unfiltered information, he could be persuaded to work with the District on the St. Petersburg permits.

Already St. Petersburg was showing signs that it didn't want to be wed entirely to Pinellas County. A March 21, 1995, article in the *St. Petersburg Times* highlighted a growing division between the county and its biggest city. St. Petersburg city councillor Paul Yingst, still sitting on the Authority board, was quoted in a weekend story that St. Petersburg could live with the cutbacks imposed by the District. Pinellas commissioner Rainey, unaccustomed to being challenged on water issues, threatened to withhold support on the construction of St. Petersburg's Thunderdome (later called Tropicana Field)—a step toward a new baseball team and revitalization of the community.

"You forget what county you live in; I'm going to forget what city you live in," Rainey was quoted as saying.

Challenged publicly at the next Authority meeting, Yingst backed down. The Pinellas message to St. Petersburg was clear: "Get in line. Now." In the *Times* article, Pinellas commissioner Robert "Bob" Stewart, who had spent years in St. Petersburg politics, noted, "That's hardball politics . . . and the water issue is the epitome of hardball politics."

We called it hydropolitics, a term first used in 1979 in a book about water in the Nile region. It's widely used today.

"After that meeting, I called St. Petersburg mayor Dave Fischer and said, 'You have to come on this board,'" Hillsborough commissioner and Authority chair Ed Turanchik remembered. "Before he came onto the board, I spent a lot of time with him trying to work on the issues; Tampa mayor [Dick] Greco, too." Turanchik felt strongly that if Greco and Fischer would engage on the issue, it could be resolved. Greco refused. According to his staff, he knew the importance of water; he just thought he could be more helpful outside the board. It was the same strategy he had used in the 1970s.

District staff saw this public divide between Pinellas and St. Petersburg as an indication that their communication strategy was working. St. Petersburg really did represent good, reasonable policies, the District thought. St. Petersburg didn't really want to be wedded to Pinellas and its bullying tactics. This public split showed that tensions were growing. The Coordinating Council had the potential to engage new people from those separate government bodies, to create new avenues of communication.

A specific program of presentations was developed to develop a shared body of knowledge to the Coordinating Council. Meetings were held in different locations and open to the public. When possible, the presentations and the discussions were televised. It was at a Coordinating Council meeting that the District executives first met the Authority's new general manager.

***

When Jerry Maxwell was a boy, he and his parents would pack into their car in the Midwest and head to Florida to visit Uncle Tommy. Uncle Tommy was a test pilot, the exciting, derring-do kind of uncle that every little boy wants. Uncle Tommy had a place on the Gulf Coast, north of Pasco County. Jerry remembered the feeling of standing on the beach, looking this way and that, seeing nothing but blue sky, sandy beaches, and fiddler crabs. Jerry Maxwell remembered the Florida of his childhood. It doesn't look like that anymore. "I am an environmentalist," he insisted, using the story to stake his flag in the world of conservationists. As it turned out, few believed him.

***

Jerry Maxwell got a call from an old friend—an executive recruiter—but he didn't much think about it at first.

"He asked me to apply for the Authority's general-manager position," he said. "I wasn't much interested in moving again, but he was persistent, so I sent in a package. It's common for a government administrator to move from place to place seeking new challenges or better opportunities."

Maxwell had worked in Lee and Alachua Counties and for Tallahassee. "Everyone in the state knew about the problems in Tampa Bay," Maxwell said. "This job would be a huge challenge." The interview process, both formal and informal, included the usual questions and answers from Authority board members and selected others. They avoided the Sunshine Law by holding individual, private interviews with the candidate. It was an unusual move but not illegal. Pinellas commissioner Chuck Rainey and Pinellas utilities director Pick Talley met with him twice.

Maxwell counted among his best qualities being "goal driven," and there were many people who believed it would take an inordinate amount of focus to gain the cooperation of local governments and bring new water supplies to the region. Maxwell said he was determined to be his own person, to avoid political alliances. In interviews with Chuck Rainey and discussions with Pick Talley, Maxwell was sure the message got through.

"I heard from someone that Commissioner Rainey said he didn't think he could control me," Jerry said, smiling. Given Rainey's political power, tenure, and history of controlling the Authority and its general manager, it seemed Maxwell was staking out personal space.

Gilliam Clarke said this: "Turanchik told me that Rainey hated him, so Turanchik loved him. I couldn't believe he fell for that old trick."

As with most people who've been in public service for any length of time, Jerry Maxwell's personality and management style were easy to document. Beyond his résumé, everyone knows someone who knows someone who worked somewhere who has information about how he works, what he's like, and whether he will fit in. But Maxwell was different.

"He was a little hard to pin down," Roy Harrell observed. "He was obviously a strong-minded individual, and they surely needed that over there, but I had the impression that Jerry had his own policy in mind."

Pasco County commissioner Ann Hildebrand was chair of the Authority as the selection process proceeded. A member of the Authority board since 1990, Ann was there as the District and Authority moved from cooperation to conflict. One of the few women in the battle and as a Pasco commissioner, Ann was different from the others. She was attacked by activists for her long-term view of the water-supply problem. Ann believed slow and steady would win the race. Back then she believed the best hope for Pasco was to get the Authority to agree on a water-supply plan, then work the plan. In her mind, she wanted a general manager who would work the plan. In Maxwell's mind, the charge for his new job was simple: "Make water work." How he did it would become its own issue and transform him into the water supplier that everyone loved to hate.

"He was Mr. GQ," said Hildebrand. "When he walked in, he commanded attention, very well put together; and he is very, very articulate. He was smooth."

In the spring of 1995, Jerry Maxwell attended his first meeting as the general manager of the Authority. Bruce Kennedy, the interim general manager, never had a chance to win the job. Rainey used anything and everything to demonstrate his disdain for Kennedy. Some agreed that Bruce Kennedy was too willing to work with the District, something Commissioner Rainey couldn't stand. Others suggested that Rainey needed someone more fully committed to Pinellas. After all, in the view of many, the Authority had been a creation of Pinellas. By all appearances, the Authority management and Pinellas government seemed to share the same policy viewpoint. It was a strategy that had worked since the Authority's establishment. At least it worked for Pinellas County.

Jerry Maxwell's first public appearance as the general manager was before the Coordinating Council. He had the look of senior management. Athletic, attractive, and with exquisitely coiffed silver hair, Maxwell exuded an air of confidence that annoyed some people. As with Steve Monsees's presentation to the Governing Board the year before, everyone there remembered that first encounter with him. The meeting was held in the Pinellas County Commission chambers. It's an odd-shaped room. In most government boardrooms, the board is central and up front. In this room, the Pinellas County Commission sits to the right from the audience's perspective. Some of the spectator seats face the wall.

District executive director Peter Hubbell remembers meeting Jerry Maxwell. "He blew into the room as we were trying to get prepared and started giving me a hard time about 'what's going to go on tonight.'"

"Why are you asking me this?" Pete responded. "We're going to go in there in five minutes and do what the hell we need to do." When Jerry left, Hubbell turned to the facilitator. "This looks like a problem."

Maxwell remembered that first meeting, but for the life of him, he couldn't figure out why the council existed. "It was like a shadow board," he said. "I mean, what was the point? Was it some elaborate PR plan? Something for visibility? I just didn't get it."

"Our goal was to bump up the discussion a bit," said Hubbell. "To get it away from utility directors and narrowly defined technicians. We wanted to look at long-term sustainable water supplies, and we wanted that discussion outside of the Authority." Besides, Pinellas commissioner Bruce Tyndall had suggested the idea. The District had simply obliged.

District assistant executive director Mark Farrell remembered meeting Jerry there, too. "He came up and said, 'You're the son of a bitch that started all this desalination stuff, aren't you? Well, we'll have desal over my dead body.'"

"I think I said something along those lines," Jerry remembered, smiling. He denied the part about "over my dead body" but not stridently. Desalination was the District's big plan for a sustainable water supply. Pinellas County's board was dead set against it, as were several Authority board members.

Based on what Maxwell said at that first public meeting, Governing Board member Roy Harrell's initial evaluation of Maxwell's potential for independent action changed. "The Authority fought us over alternative supplies," Roy said. "At that first meeting, Jerry came in pretty much espousing Chuck's policy: groundwater, groundwater, groundwater, no question about it. Chuck Rainey was calling the shots."

It's unusual for someone to walk into a first meeting at a new job and begin sharing opinions as Maxwell did. It wasn't just that the District disagreed with his opinions; it was that he was so new to the argument. How was it possible to hold such strong opinions already?

Hubbell agreed with Harrell's assessment. "It looked like he'd been given the script by Pinellas, and Pinellas was coming in strong. My biggest issue with Jerry was that he had the opportunity to come in, take an honest, objective look at what's going on, and do things differently instead of being controlled by the forces that be—Pinellas," Hubbell said, reflecting on the tumultuous past few years. "He came in and became their boy."

Still, Hubbell admitted that the job Maxwell took over was indeed a tough one. Member governments were fighting among themselves, the environmental damage was in the media nearly daily, the utility directors were circumventing opportunities to strengthen the Authority, the District wasn't giving much leeway, and a myriad of other problems were evident.

The Tampa Bay Water Coordinating Council continued to meet for several months. It had a professional facilitator who was paid by all parties. It was listening to presentations on everything from water usage to alternative supplies, from conservation to costs. District staff felt like they were making headway. They thought their arguments were better and more powerful than those of Pinellas County or Authority staff.

"One day before a meeting, Pinellas commissioner Bruce Tyndall came up to Pete and me," Joe Davis remembered, "and said, 'Maybe we should be listening to what the water-management district is saying on some of these issues, and not our staff.' Pete and I just kind of looked at each other in disbelief."

Coming from one of the hardliners who wanted only groundwater, it was either an amazing shift or a blatant political maneuver. In either case, many meetings were held, and much technical material was covered. For elected officials, understanding the minutiae of a problem is not easy. They must rely on their staff specialist(s) to guide them. The problems are too big, complex, and numerous to understand each one in depth. The water issue required elected officials to commit to a level of understanding to which they were unaccustomed.

"Water is complicated stuff," St. Petersburg mayor Fischer said. "It's complicated chemically, it's complicated environmentally, and it's complicated politically." He pointed out that there were competing ideas about what was happening in water supply policy. "There were three theories," he said. "We've got all the groundwater in the world, we told you so, and you're doing environmental damage."

The Pinellas approach said that fifty-four inches of rain annually was plenty. The St. Petersburg approach was "We told you our system is in balance. If we don't increase pumping, we're not going to have a problem." The District's point of view was "There's damage, and something must be done." As the District began to talk about alternative supplies they hoped there might be some compromise.

"Mayor Fischer got away from his utility director, Bill Johnson," Mark Farrell thought. "He was getting real information now, unfiltered information, and he was learning directly about the water resource and the options."

But the issue wasn't just Bill Johnson—it was all the utility directors. The Authority's Technical Advisory Committee consisted of its utility directors, who were supposed to make recommendations to the Authority on matters affecting shared water supplies. But in practice, water wasn't really shared. It was owned separately by the local governments or in cooperation with the Authority. The utility directors had their permits (or permits pending) lawfully issued by the District; they had sufficient capacity to meet planned demand. To them, there was no need for a change. Change could affect water-treatment processes, change could affect rates, change could alter the way things were, and things were just fine, so why change?

But policy makers were faced with a different set of issues. Policy makers had to worry about intergovernmental relations, public perception, impacts on people or the environment, and the future, in addition to added costs.

"It was cheaper to litigate than to add infrastructure," Mark Farrell said. "So the litigation continued, and no one would admit publicly that there were real impacts—except the District."

But at the Coordinating Council meetings, the impacts, the alternatives, and the costs were under considerable scrutiny.

"We got the lawyers out of the way, we got the TAC [Technical Advisory Committee] out of the way, and we could see things beginning to change a little," said Farrell.

The media showed marginal attention to the meetings. There was a lot of technical information, and it was not a place for public input, although activists regularly attended. The meetings were structured for information and discussion.

"I would say to Pinellas County, 'All this litigation has not given us one drop of water,'" said Governing Board member Roy Harrell. "They would say, 'Yeah, but it hasn't gotten one drop of cutback either.' When I heard that, I knew we would litigate forever."

The Coordinating Council continued to meet and even produced a final report that was never adopted. Pinellas commissioner Bruce Tyndall refused to set a date for a review of the draft report, and the committee's work just petered out. Then the District made a serious mistake. The final report was used as an exhibit in the pending litigation—something all sides had agreed they wouldn't do. At best, it was an error; at worst, it was a breach of good faith. As the executive working with the legal team, Farrell took responsibility.

"That was a low point," Governing Board chairman Joe Davis said. "We had put so much time and effort into building those relationships, having those discussions, and then we did what we said we wouldn't do, and everything fell apart."

While mostly uninterested in the meetings themselves, the media paid considerable attention when informed that the District had erred. Publicly accused by Pinellas of breaching the agreement establishing the Coordinating Council, the District admitted the error and countered that Pinellas was violating the agreement, which promised not to make public statements that could prejudice the council's decisions. No matter. Joe Davis apologized, the report was withdrawn from evidence, and the council folded without ever completing its work. It was a huge mistake. And, as a result, no one will never know whether the council might have produced a joint resolution to a shared problem.

As the Coordinating Council ended, litigation and negotiation continued over the wellfield permits, the water-shortage orders, and the emergency order. Through 1995, the District considered recalling the other permits in the central system but hesitated. Eventually, they would all be in for review.

The District kept hoping that agreement could be reached, while St. Petersburg kept hoping that political changes would force the issue. When Governor Lawton Chiles was reelected in 1994, narrowly beating challenger Jeb Bush, all the water-management districts breathed a sigh of relief.

Mark Farrell captured the sense at the time: "The Republicans had taken both houses in Tallahassee. If Chiles lost, we would have had to start all over, if we could start at all. No. If we were going to change things, it was now."

"We never played the legal gamesmanship that the other side did," Executive Director Hubbell claimed. "We were strategic, and they were tactical. They did everything they could to slow us down. For them, it was the politics of delay. If they could delay policy change long enough, the politics might change."

"Ed de la Parte is a great lawyer," said St. Petersburg mayor Fischer. "He can pile on the facts and the figures. My staff admired his work, and if we were going to battle, that's what they wanted," he recalled. "But when I thought about it, he could keep you in court for twenty years until the wellfields were shot. If we tied this up in court for another twenty years, we wouldn't have any more water. We've got to build a system." He was convinced that litigation was not the way to do it.

"But the litigation pushed people toward trying to find a solution," Hillsborough commissioner Turanchik said.

Here again, two people see the same situation differently.

"We were on the right side of the issues," Hubbell said. "We had a commonsense approach, and we explained that to the public. That's why we had so much public support. We made sense. They didn't."

The legal bills dragged on for all the involved governments and some of the residents. Most observers credit Pinellas County and its lawyer Ed de la Parte with raising the stakes just as Mayor Fischer feared. Once the Authority hired a staff general counsel, de la Parte was temporarily out of the water wars. Some say his departure from the Authority was due to media coverage of his high fees. More likely, when the new general counsel started, he wanted a fresh approach. He couldn't do that with de la Parte as a visible actor. Don Conn, a former administrative-hearing officer, became the general counsel for the Authority, and de la Parte was invited to represent Pinellas County. The aggressive Pinellas Commission, coupled with the talented and aggressive de la Parte, put the District on the defensive—legally. Publicly, the Pinellas water lawyer would eventually become an issue.

"Eddie was incredible," said Jerry Maxwell. "He took the smallest item and made a huge deal out of it legally and legitimately. Does it serve billing purposes? Oh yes. But these are legitimate legal issues, so why not pursue them if the client is willing?"

"Ed drives up the cost of litigation by filing every motion that can be filed, holding every deposition that can be held, and I think that's why he was considered part of the problem," said water lawyer Doug Manson, who was watching the legal action unfold. He was at the District when Steve Monsees gave his persuasive presentation. Things had obviously escalated since.

While Hubbell agreed that de la Parte may be a good lawyer technically, he saw the tactics as a reflection of personal values. He didn't approve.

"Ed thinks very short term," Hubbell said of de la Parte. "It was frustrating. We had a lot on our plate, and all he did was push the issues in Tampa Bay. He billed nineteen hours a day; that's all he did."

Hubbell believed that the biggest impediment to a regional solution was Ed de la Parte. "He was laughing all the way to the bank. He was acting way beyond an attorney," Hubbell said. "He was guiding public policy."

Many agreed with Hubbell. In the coming year, though, de la Parte's advice would critically damage Pinellas County. With Pinellas and de la Parte leading the challenge to the District's authority, the staff knew that they couldn't afford another misstep. Pinellas used its legislators to distract

District executives with charges and investigations of alleged misdeeds. Even so, the District continued to coordinate with the Department of Environmental Protection and the governor's office.

"Every step of the way, we were in close touch with the governor's office," Hubbell said. "We worked directly with the lieutenant governor. We'd say, 'Buddy [McKay], here's what we're going to do,' and give it the gasp test. You know, how deep a breath did he take when he heard it?" Hubble chuckled, thinking of the hyperventilating that might have been going on in Tallahassee based on the District's actions. "They stayed strong, though. The lieutenant governor was on the right side of the issue."

Though it pressed ahead with aggressive litigation, Pinellas felt it couldn't get a fair public hearing. The county was being pounded by the two daily newspapers and by television and radio. The *St. Petersburg Times* was especially harsh, and it was the hometown paper.

"I called them for a visit," Pinellas commissioner Steve Seibert said. "Until that time, I'd never talked to the *Times* about water. I thought it was irresponsible of them to write their editorials without seeking my opinion. On water, they talked to one of two people, Pick or Chuck." But Seibert said that even when he talked to the editorial-board members and was his most charming and reasonable, two days later an editorial blasted the commission for its aggressive tactics. It couldn't get its story out, he said.

It's possible no one was buying its story. Some would say that Ed de la Parte behaved unethically as he threw one legal challenge after another at the District. Others would say he was zealously representing his client. As with most choices in the story, it depends on perspective. It wasn't de la Parte's job to concern himself with the region. He was hired to defend the status quo, and that's what he did. The District's use of a report in litigation was inadvertent, they claimed, but does it matter? The end result was the same: tentative trust was shattered, but the value of shared facts was tangible.

Chapter 12
The Rising Tide: A New Policy

Judy Williams and her husband dreamed of a beautiful home on a lake where they could water-ski. They searched and searched until they found a three-bedroom, two-bath house on Daiquiri Lane in central Pasco County. It wasn't perfect when they saw it, but before they moved in, they added a second floor, enclosed the carport, and added a boat ramp, even though the boat, at the time, was wishful thinking. The District calculated Bird Lake to be 150 acres, but most of that was wetland, according to Judy. In any case, when renovations were complete, the young family moved in. One six-year-old and one newborn; a girl and a boy.

"I wanted my kids to grow up on the water," Judy said.

To complete their family, they visited the Humane Society and rescued a golden/lab dog. "Lady was a pale, golden color," Williams remembered. "So sweet."

Life was sweet.

Then Judy noticed the lake level dropping. She didn't think about it at first; Florida has periodic droughts punctuated and balanced with drenching rains. But this seemed different. And as the wetlands began to shrink, Judy noticed one other thing. The alligators that lived there were increasingly seen along the shore. She talked with the neighbors who told her that the lake had never been so low.

One spring evening, she was heading out the door to a meeting. "Don't forget to bring Lady in," Judy told her daughter. When she came home a few hours later, daughter Sarah was distraught. "Lady didn't come when called," she fretted. "She always came." The next day flyers went up and neighbors were contacted. A few days later, one had news.

"I saw a huge gator dragging a dog," the woman told Judy. Judy recalled, "I was so mad, I got my canoe and followed it, but it was too late. The dog was dead. It was Lady."

By the time Lady was taken, the youngest of the Williams family was eighteen months old. He loved the water just as his mother hoped.

"It could have been my toddler," Judy thought. "No. This has got to change."

***

Just before the close of the 1995 legislative session, three citizen groups representing many counties and constituencies—Taxpayers Outrage Organization for Accountable Representation (TOOFAR), Worried Water Watchers, and COLA—chartered ten buses to Tallahassee to address state lawmakers. They were calling for change from the present (and failed, they said) water policy. The six hundred people were residents of Citrus, Pasco, and Hernando Counties, and their message was clear: "Don't suck us dry to provide water for Tampa Bay." Of those counties, only Pasco was in Tampa Bay. Hernando and Citrus, like Pasco before them, were still rural. Both had plenty of water they intended to keep. They didn't want Pinellas or the Authority buying tracts of land to harvest water to send south. On May 5, 1995, the *St. Petersburg Times* trumpeted their message: "Desal Now" and "Local Sources First." The climate had clearly changed.

Mark Farrell summed it up neatly. "We had a charged-up board who wanted to do the right thing. We had the technical information to do what we needed to do, and we had the guts to do it." It was a recipe for incredible success—or total failure.

As negotiations on the litigation dragged on, it was abundantly clear that there were two basic philosophies at work. In the first, advocates suggested if we didn't have enough water in our own backyard, we should push our pipes and wellfields further out to get the cheap water from the next location, or the next. After all, the rationale goes, in Florida, water belongs to everyone, so we have a right to it. Opponents asked, "Why should those people be willing to share their water with us? Look at the wellfields. Have we been good stewards? Why should other communities sacrifice their future growth while we take their cheap water to fuel ours?" Feelings were intense on both sides.

As a matter of policy, the District wanted to establish some drought-proof water supplies. Many people believed that extending the pipelines just moved the problem to another rainfall-dependent water source. Like the New Water Sources Initiative, Local Sources First became a passion for District chairman Joe Davis. Whatever divide existed when Davis was appointed to the board was long gone. He came to the board with a mindset to cut waste, and cut he did. "We don't have a water-shortage problem; we have a water-delivery problem" was the message from the Authority, Pinellas, and St. Petersburg. The District didn't think so. There was a water-supply-limit problem, a cooperation problem, and a failure to admit any environmental impacts. To prevail, the District needed to demonstrate what was wrong with framing the issue as a distribution problem.

Local Sources First became a District focus. It had great appeal. The policy didn't say "Local Sources Only." It merely said that a community should exhaust all its available water supplies—including technologies like desalination and aquifer storage and recovery—before turning to other communities for their surface water or groundwater. The values embedded in such a notion tapped into core values for most Americans. Local Sources First suggested that we should draw on our own resources before we reach for someone else's. It was not so much that the District painted other ideas in a bad light; it was just that Local Sources First was so sensible, so logical, and so abundantly fair that the alternatives seemed wrong by contrast.

"Yes, the ownership of water is very antienvironmental," said Doug Manson, sarcastically. "What Local Sources First really suggests is that counties somehow have a right to the water under their land. What's wrong with that?" That's not the way the law worked, though.

An op-ed by Chairman Davis was distributed district-wide. The column was the most widely run piece of all the materials produced during the water wars. It had face validity, it tapped into cultural value of self-sufficiency, and it didn't rule out other possibilities. Local Sources First suggested that we should treat our neighbors the way we wanted to be treated. Florida law allowed water to be taken from elsewhere. Although Florida law permitted the taking, was it right, ethical, or proper? The people who lived near the wellfields in Hillsborough and Pasco Counties didn't think so, and neither did the District. The District wanted to ensure that any future mistakes would not look like past mistakes.

Local Sources First had many supporters. The release of the op-ed brought more than coverage. It brought out supporters, the kind that read about issues in the paper or see them on TV and decide what makes sense and what doesn't. The supporters who came to Local Sources First were not activists. They were not people directly affected by wellfield pumping. They were not people with a specific environmental issue related to the overuse of existing water resources. They were just people who thought that it was fair to take care of yourself before asking others to do it for you. That message moved to a new audience and to a new, broader constituency.

During the 1995 legislative session, the District drew a lot of attention. Actions in the south (the SWUCA), in the north (removal of the controversial Wysong Dam), and in the middle (engagement in the water wars) were drawing attention. The Wysong Dam was an inflatable structure on the Withlacoochee River removed in the 1980s. Hubbell and Farrell, both of whom preferred nonstructural water control, had indicated that it was no longer working. The activist group TOOFAR had formed in response. They wanted their dam back, and they wanted the District to pay for it. "Over my dead body" was Hubbell's response. As with all water issues, the fights were emotional and tense. At one public meeting, a man shouting at Hubbell had a heart attack and died right there on the floor.

TOOFAR supported the District on alternative sources but opposed it on virtually everything else. At community events, the group would distribute District literature on one side of their booth while they handed out "Ditch Hubbell" bumper stickers on the other. Hubbell kept that sticker on display in his office until he left the District.

Depending on which side of an issue you found yourself, the board and District leadership were either making up for lost time by restoring the environment and putting some management into water management, or they were out-of-control bureaucrats—exercising power they didn't have to accomplish tasks that didn't need to be done. No matter; the District was taking political heat from the Florida legislature and the other water-management districts. The District commissioned another public-opinion survey. This time, respondents were more directly asked about transporting water. The public supported Local Sources First and the District's actions to put the policy in place.

In 1995, public support was solid. In a survey conducted by Market Insight, this time including eight of sixteen counties, the District still had 74 percent support for the job it was doing. It was a pleasant surprise. The water issue had transformed from a protest by affected parties into a political movement supporting a specific policy.

"There was political pressure building," Mayor Fischer remembered. "And there was definitely a movement."

"We took a beating from the other districts," Hubbell observed. "All this stuff that was for Tampa Bay, and it might become state law. We were too visible. The other districts were worried that we were taking so much heat, the institution of water-management districts might be affected."

Local Sources First was a good policy, Hubbell maintained, and it was so fair that it was hard to argue against it (although Pinellas County did). Water lawyer Doug Manson said it wasn't hard to argue against it, but St. Pete, Pinellas and the Authority just couldn't get any media coverage for their argument. At one point Pinellas commissioner Rainey was quoted in the *Tallahassee Democrat* newspaper as saying, "Keep the Suwanee River cold, because we're coming to get it." It was typical Rainey, but few were laughing.

"He was the District's greatest ally," Doug Manson said. "His complete failure to address public opinion gave the District the enemy it had to have.

It was good versus evil." Manson's future client, St. Petersburg, was standing shoulder to shoulder with "evil."

Making Waves: Investigations and Distractions

In response to pressure to change water-supply policy, the pressure to leave it alone increased. The District was moving to prevent or limit long-distance transfers of water in its region. But powerful forces throughout the state liked water policy just the way it was. With the Tampa Bay–area problem dominating statewide water news, the region's crisis was driving state water policy. While District executives worked closely with the governor's office, state legislators (particularly in West Central Florida) were hearing from their constituents. No water-management district had exercised this kind of authority before. No water-management district had ever challenged a powerful local government like Pinellas County. And even though there was a Democrat in the statehouse, both houses of the legislature were controlled by Republicans. Even the other districts wanted the District to back off—a little bit.

The Water Management District Review Commission was established to reexamine both the system of water management in Florida and the individual districts. It was an opportunity for critics and supporters to voice their hopes, concerns, frustrations, and observations to a panel of political appointees. Many of the members were legislators, and a few represented special interests like real estate, the Audubon Society, or agriculture. Some said that Senator John McKay of Sarasota pushed hard for the commission. He believed that the District (embroiled in the Tampa Bay water wars and conflict regarding overallocation in the Sarasota region) was out of control, and he wanted to bring it back into line. Meetings were held all over the state. Each district had its supporters and detractors.

***

I attended a meeting of the communication directors for the districts. Jerry Scarborough, the Suwanee River Water Management District executive director, presided over this subcommittee. The meeting was held in Palatka, headquarters for the St. Johns River Water Management District.

It was much more like the Taj Mahal than our Brooksville office was, I thought, and in a smaller community. The meeting began quite cordially. Scarborough called on each of the communication directors to talk about what they were working on with the Water Management Review Commission and otherwise. Finally, he called on me. Whatever I said triggered him. He raised his voice, pointed at me, and loudly said, "This is your fault!"

"My fault?"

"You people got this going," he declared. He was furious.

"We have a lot of support," I told him. "Look at our public-opinion numbers." I loved showing those off.

"I don't care." He glared at me. "This review commission is a problem, and it's your fault. You guys better get a grip. This is a problem for everyone." Or words to that effect.

His reaction came from the fact that we confronted our water problem, I think. Districts just didn't challenge permittees like that—especially local governments. During a break, I called the office and told Mark Farrell what happened. "If it's a problem, leave," he told me. "You belong to us. He shouldn't be yelling at you."

"Oh, only you guys can yell at me?" I asked, joking.

"Yeah," Mark said. "I'll talk to Jerry. No big thing."

"No big thing," I agreed.

Except it gave me a taste of what Mark and Pete and the Governing Board had been getting from the legislature, around the state, and from their colleagues in the other districts. My significant contacts had been activists and the media. I had a new appreciation of what they were personally sacrificing to confront our environmental problems.

***

"When someone is trying to divert your attention, when they're trying to distract you from the job you're trying to do, they review you, they audit you, they play games with you," Mark Farrell said. But neither the staff nor the Governing Board gave in.

Governor Chiles, who appointed the Water Management District Review Commission, was not trying to disrupt with the districts. He supported the policy changes that were evolving. Maybe the Review Commission was a way to keep the legislature from acting, perhaps it was a delay tactic, and perhaps it was in response to powerful agricultural forces that resisted what they saw as increased District authority. Whatever the reasons, the selection of members for the committee would determine its potential outcome. All the districts lobbied to have "friendlies" on it. Despite their best efforts, Governor Chiles appointed some harsh critics, mostly agricultural interests. But one key Chiles appointment was former Florida senator Phil Lewis. Lewis, who chaired the Review Commission, was one of the original architects of the district system.

For about eighteen months, the commission traveled around the state taking public comment, hearing presentations from the water-management districts and about them, and asking a lot of questions. Peter Hubbell, Mark Farrell, a handful of staff, and sometimes Governing Board members attended. Preparation for the meetings took huge amounts of time, none of which included the personal visits and phone calls between meetings and educating those Review Commission members who either were inclined to support the District or had an open mind. To some extent, the districts tried to coordinate their efforts, but each was so different it was hard to do. The issues for Southwest were and are different from South Florida, where they were embroiled with Everglades challenges. Northwest and Suwanee needed more money, and St. Johns had water-quality issues. Each district had its critics and its supporters.

***

One Review Commission meeting was held in Bradenton, in the Southwest District, to discuss education. I decided to go. Ironically, other than identifying statutory mandates for education and communication, we did little preparation. We let our education advisory committee know that the meeting was taking place.

Because the District's communication program had been successful, it was under attack. Already, my face had been on the front page of the *Tampa Tribune* in an article that detailed our "public relations" expenditures. In fact, more than half of the $3 million spent on public communication was spent on programs for schoolchildren. Those programs emphasized watershed education, conservation, and science. Another large portion of the budget was allocated to public education, much of which was for neighborhoods and community grants. What critics hated was our public-information effort. That consisted of a media specialist and me. What made them crazy was that all our education efforts were narrowly focused on the water-resource issue and related issues. We were solidly engaged in education and information that was on mission and mandated by statute, and that made it difficult for critics to point their finger at something that was wrong or bad. The program was designed that way. I knew when I was hired that if the communication program was effective, it would come under attack. I wanted to make sure we could defend it. We had the opportunity that night in Bradenton.

State senator John McKay (who later became the Senate president) was particularly brutal. He held up newspaper articles and criticized the executive director personally, pointedly, and publicly. I stood in the back of the room, hands clenched, teeth gritted against the things he was saying and the way he was doing it. Not everyone attacked, but it was singularly unpleasant.

When asked to respond, Hubbell leaned into his microphone and said, "I am *not* going to apologize for our education program. Children must understand the water resource, and people have a right to know."

I know that his answer was longer and more detailed, and I know—because I was there—that he said it in terms that were more diplomatic. But he did say, "I'm not going to apologize" in precisely those words. When he said it, I felt a warm rush. No matter what Peter Hubbell ever said or did, I will always remember that moment and his comment and know what he is fundamentally made of. At that moment, I loved Pete Hubbell more than I can say.

***

In 1996, the Review Commission issued a report of its findings. The bottom line: the water-management districts, while far from perfect, have a big, tough job to do and are doing it reasonably well. The big, threatened reorganization of water regulation in Florida never materialized—until decades later when then governor Rick Scott essentially removed the top three levels of management in each District across the state, decapitating them and removing institutional knowledge and experience. Perhaps the commission was meant as a political distraction, but the districts—all of them—had weathered it well.

In Tampa, negotiation and litigation continued, but nothing was resolved. It seemed as if the parties in Tampa Bay would rather have a trial judge set water policy than work it out themselves.

Make Water Work

While the District was engaged with defending itself and advancing alternative-water-source policies, the Authority was undergoing its own changes. True to his charge, new general manager Jerry Maxwell was focused on making water work. "I was focused on finding new water supplies and trying to get the member governments to support them," he said.

In the summer of 1995, District executive director Pete Hubbell and chairman Joe Davis had several meetings with Senator Jack Latvala, whose constituency covered Pinellas County and part of Pasco County. Latvala, known for his hard stances and willingness to fight the good fight, listened carefully. Policy makers and others who work with him have a saying: "Latvala is Latvala," meaning that you must deal with his blustering to get to the issue, but if you're able to get him on your side, it's worth it.

"In one of the meetings, Joe and I sat there talking about the need for a strong central utility," Hubbell said. "Pick Talley was in the room whining about Florida getting fifty-two to fifty-four inches of rain every year and how we had more than we could ever use." Hubbell shook his head, partly disgusted and partly sad because Talley failed to realize that things were changing around him. "Latvala kept looking at Pick as if to say, 'What the hell are you talking about?' Pick had this lame thermometer with the fifty-four inches on it, and I think Latvala must have been thinking, 'Give me something to go on here' because Pinellas wasn't looking very good at all. Joe and I knew that we were making sense."

Latvala was interested in strengthening the Authority if that would end the litigation and conflict. Supported by Hillsborough and Pasco Counties and with St. Petersburg in serious permit negotiations, Latvala had something to work with. He had the credibility and power to broker a deal. He seemed determined to do it, although some advisers told him it was a no-win situation.

"That's the thing about Latvala," said St. Petersburg mayor Fischer. "When he's ready to get the job done, he commits completely." Latvala had one thing no other legislator had: as chair of the powerful Senate Natural Resources Committee and a ranking Republican, he could back up threats of legislative action with legislation. For the first time, someone outside of Pinellas's reach and with tremendous political power doubted the soundness of the Pinellas position.

In the meantime, the Authority was struggling to identify sources of new water. Member governments, already suffering from present overuse, were reluctant to offer up new sources within their boundaries. At the same time, the Authority was under pressure from business, the media, and the public to adopt a plan for future water supplies. Desalination was gaining favor, as were reuse, the rehydration of wetlands, and even the treatment of wastewater and its placement in the potable-water system. Pinellas County was appalled. It was losing ground on every front.

"In the long run, the big loser in the water wars was Pinellas County," observed Joe Davis. "Not just because of the loss of groundwater, but because of the loss of political prestige and influence."

But Pinellas's biggest mistakes were yet to come.

"Councilmember Paul Yingst went to the city-council meetings and told us how bitter the fight was," Mayor Fischer said. "Our utility director, our lawyers, everyone was hunkering down for a long battle. So when Paul was defeated in the election, I appointed myself to the Authority board. I wanted to see for myself why there was a deadlock, what the personalities were. There were heavy legal bills, no new water, and no new plans. I knew something had to change."

After Mayor Fischer appointed himself to the Authority representing St. Petersburg, Ed Collins, the new chair of the Pasco County Commission, appointed himself as its representative. "It was time for change," he said. Gilliam and the water activists applauded the decision. Hildebrand was off, and Collins was on.

By the end of 1995, there would be a significant step forward. The Authority finally adopted a water-supply plan. In some ways, the District was pleased. It was better than no plan. On the other hand, developing alternative water supplies was not a high priority. In that, the District was disappointed. Still, a plan could be modified, and over the next two years, that one was—repeatedly.

Negotiate—but Prepare for Trial

Doug Manson had been watching the water wars from the sidelines. Critical of the District in the Review Commission hearings, he'd been speaking on behalf of his agricultural clients. Still, he was present at Governing Board meetings and heard the arguments of both sides.

"I hadn't worked for public supply much," Manson admitted, "but I knew the environmental impact, and I knew water law." Sitting through the meetings, he'd often chat with Kim Streeter, St. Petersburg's staff counsel, about strategy and tactics. When outside counsel Kent Zaiser had to step aside for another case, Streeter recommended Manson for the job of outside counsel working with Dan Fernandez, the former District lawyer.

Yet again, District staff were pleased. Doug seemed a good fit. He was the kind of lawyer you could get a settlement with, the kind you could work with. They had done it before on agricultural issues. He just might help reach a settlement to avoid litigation. It wouldn't be easy, but at least it was possible.

"I joined in the fall of '95," said Manson. "The case was on its way. The dynamic among the parties was already there. The trial was set for January, so I spent most of my time preparing for that. But the District was doing this weird polka dance with court dates, so the case ended up being delayed twice more, until September of 1996."

When Manson joined the St. Petersburg team, the first thing he told them was "I cannot take the position that there are no impacts, because it is obvious there have been changes in character of the wellfields. But no one in St. Petersburg was saying that," according to Manson. "Those wellfields had been there for thirty years. Whatever environmental impact they had, was done. There was a price to be paid, and you have to weigh that against the benefit."

By the time Manson entered the fray, the experts had already been selected and the strategy set. It was his job to figure out what kind of information was available so that he could know what kind of story he could tell—and prove.

"Normally I'll ask the expert, 'What do you think the case is?' and let them tell me. That way I know what they can comfortably say—what they believe—and we go back and forth until we have a story that they really believe so that they can be a good witness," he explained.

The relationship among Pinellas, the Authority, and St. Petersburg was strange, Doug said. "Ed de la Parte would put forth strategies, and we'd discuss them, accept some, reject others. We worked closely with Pinellas, less with the Authority. Pinellas and Ed de la Parte's support were great because they had everything from the beginning of time dealing with this case—already notebooked and cataloged. It was a huge help."

Although de la Parte had been involved in discussions, it was March before he participated in depositions. Until March, Manson explained, de la Parte would have been in his office categorizing, cross-referencing, labeling, summarizing, and indexing all his work. "When that was through, he was ready for depositions," Manson said. "Until then, other lawyers handled depositions on behalf of Pinellas. We never saw him because he was busy."

De la Parte's rule for the trial came from a mentor—attorney John Allen—who said, "You can overtry a case, but you can't overprepare." Manson's evaluation of de la Parte's work ethic matched the District's. They said de la Parte was amazing when he was on script, but if you gave him some surprises, he didn't do as well. The District lawyers wanted to give him some surprises during the trial to get him off script, to gain some advantage. In the spring, they couldn't predict what the surprises would be. Once the trial began, they didn't care.

The Authority, on the other hand, was not as easy to deal with. Manson remembered that "some of the consultants were particularly hardheaded. They were trying to tell us how to try our case, and that's not how I do things. At one meeting, I remember getting into a shouting match with Allison Adams." Allison was an Authority engineer. She was not a favorite among District staff, who noted her testy personality and defensive attitude. "I told her," Manson said, "that I was tired of the Authority coming up with surprise evidence, and she said, 'You knew about this all the time,' and I said, 'No, I didn't, because you released the information piecemeal.'"

But, Manson said, by the time it came to trial, Don Conn, the general counsel for the Authority, pulled the case together. "He's a great litigator," Manson observed. "It's not that they wanted to hurt the case; it was just a lack of integration."

While the opposition was struggling to integrate its cases, the District moved ahead with its. The problem was the two sides were preparing to argue different issues. "We were like two ships passing," said Manson. "Normally, two cases collide, but there was no collision here—there were two separate cases."

St. Petersburg knew it was the test case. It had the misfortune of having its permits in for review when Steve Monsees visited the Governing Board. "Originally, the Authority's lawyer also represented St. Petersburg," Manson said, meaning Ed de la Parte. "And early on there was a contentiousness to St. Petersburg's response to the District's request for additional information."

Bill Johnson agrees. "We started out badly in 1992," he said. "When we were asked to provide information, we said no, and to this day, I think it was a mistake."

"It was like punching them in the nose to start out," Manson observed.

"We quickly changed counsel to Dan Fernandez and Kent Zaiser," said Johnson, "but it might have been too late." The adversarial relationship, in place for years, had been reinforced one more time.

In every issue there are human beings. They care about whatever it is that they are dealing with. Screaming at board members and threatening legal action and even violence is far more prevalent today. During the water wars there was occasional screaming and there were personal attacks, though they tended to be more about personality, professional competency, and positioning. To the extent that unwarranted attacks might also be unethical, lying to gain advantage certainly was. Though there was plenty of misrepresentation and more than a little strategic ambiguity, people believed that they were acting ethically. Based on what came next, the water activists would argue that some were not.

Chapter 13
The Transitional Year

SLAPPing the Activists

Eileen Hart called first.

"We've been sued," she said, even though she didn't really believe it.

"By whom?" I asked.

"Pinellas County."

"Pinellas County sued you?" I asked, unusually slow to get it. "For what?"

"Because we were gonna sue them."

"Sue who?" Still not getting it.

"Pinellas County."

***

Eileen Hart was the District's nominee for Hillsborough County's Moral Courage Award in 1996. She was a Master Gardener, wife, mother, and water activist. Eileen and her family lived in rural northwestern Hillsborough County, a region where lakes are drained, personal wells go dry, and the Authority's Good Neighbor policy is a punchline. The Good Neighbor policy says that if your well goes dry because of wellfield pumping, the Authority or the responsible member government will pay for its replacement. The activists refer to the "magic circle," an area around a wellfield within which potential impacts are presumed and the benefit of the doubt goes to the complainant. Outside the circle proving why your well went dry is an expensive and dicey challenge.

The Good Neighbor policy was debatable as public policy, but it was the existing policy. Doug Manson pointed out that the policy was only as good as the District's mitigation policy, which was woefully inadequate. Still, it was the approved operating process. In Eileen's neighborhood, in northwestern Hillsborough, a church well had gone dry. St. Petersburg resisted replacing the well, saying its wellfield was in place either before the church was built or before the church well was installed. If that were true, St. Petersburg would have precedent to not replace the well, as it was first in use. After negative media coverage, St. Petersburg replaced the church's well. Still, the experience left a bad taste with the neighbors who remember it and other encounters with the Authority, its member governments, and the Good Neighbor policy. There is a long history of broken promises and bad feelings.

Eileen Hart had defied her neighborhood association by saying that she didn't want a temporary fix to the water-supply problem, and she didn't want to fix the lakes in Hillsborough at the expense of the wetlands in Pasco. Her neighbors didn't like it, but Eileen had to do what she thought was right. Because of that, the District nominated her for Hillsborough County's Moral Courage Award. Eileen was in the water wars for the long haul. But sued by Pinellas County? Surely they couldn't be that shortsighted.

"Someone showed up at my door and handed me some papers that said, 'You are being sued, your wages can be garnished, you can be incarcerated,'" she told me. With the service came a letter from Pinellas utilities director Pick Talley stating, *"If you don't sue the County, we'll drop the charges." It was signed "Sincerely."* When God calls you to service, as Eileen believed he had, he occasionally rings the doorbell. This was more a call to arms. Eileen has three lawyers in her immediate family, so she was less worried about legal fees than about the aggravations to come.

In fact, Pinellas County had sued the District, COLA, and eight individuals. Among them were Eileen Hart, Gilliam Clarke, Judy Williams, Gaye Townsend, Bill Alexander (the school executive who had testified during the emergency-order hearing), lawyer Meredith Wester, and some of the most vocal Pinellas County critics. Wester was a District basin-board member and someone who lived in an affected community. The morning of January 17, 1996, the *St. Petersburg Times* quoted Pinellas County Commission chair Bruce Tyndall as saying, "It's something I wish we didn't have to address, but we didn't bring it up."

Ed de la Parte, representing Pinellas County, had recommended suing the activists to cut the cost of litigation. All the commissioners expressed regret at the action, and Steve Seibert opposed it. "I don't want to make this the big, bad government attacking individuals who think overpumping has damaged their lakes and wetlands," Seibert warned. "That's how it will be perceived." As a lawyer, he understood the reason for the action. As a politician, he understood the consequences. As a human, he voted for it.

Tactically, Pinellas leaders believed that the District would be forced to take their side since the District issued the permits that caused the problem. They gambled on the hunch that the District would be concerned about liability. They couldn't have been more wrong. The District was committed to fixing the problem. The citizen-activists who were sued trusted the District. They knew the only side the District would be on was theirs. Their trust was rewarded. From a strictly legal standpoint, the suit might not have been a SLAPP (strategic lawsuit against public participation), but it was interpreted that way and nothing Pinellas did could change that belief. A SLAPP suit is a tactic used by those who wish to silence opponents. Most activists cannot afford the time and money it takes to be a party to a lawsuit. And the government has bottomless pockets. Consequently, activists settle quietly while the plaintiff—typically a private interest—resumes whatever activity was the focus of activists' complaints. SLAPP suits generate a lot of bad publicity for plaintiffs. Strategically they traded bad publicity for scaring activists into silence. It's rare for a government to sue citizens for activism. The outrage felt by activists as the suit was filed was soon replaced with the horrible realization of what it's like to face the heavy hand and deep pockets of government. The elation of initial bad publicity for Pinellas gave way to the immense reality of the experience, the enormous hassle, the ongoing, unrelenting pressure, and the expense. At the same board meeting in which they decided to sue activists, Pinellas County commissioners fretted that their side of the story wasn't being told. They allocated $800,000 for a media campaign. At the time, it wasn't only the lawsuit against citizens that attracted attention. There was something else.

"They played into my hands," activist Gilliam Clarke said. She was the only person expecting the SLAPP suit, apparently. "I love Eddie. When you're a water activist, you're fighting the most powerful county in the state, and you're running on your own dime, you try to demonize them. And that only goes so far. Beyond that, you must get them to incriminate themselves, and I kept telling Judy [Williams] and some others, 'We have got to get Pinellas to show everyone what arrogant, unfeeling monsters they are.' So every time I got a chance, I would say, 'We're going to sue Pinellas County.'"

Gilliam learned about the lawsuit from a reporter. "I got a call from a *Tampa Tribune* reporter," she said. "David Pedreira said, 'Oh my God, Gilliam, I just got the news. Pinellas County has sued you.'" Gilliam grinned broadly and added, "I know he thought I was crazy, but I was glad. Finally, they had crossed the line. So I called my lawyer and said, 'Henry, Pinellas County just filed a SLAPP suit against us. What do I do now?' And he said, 'You don't worry, honey, I'll handle it.' They gave themselves the biggest black eye," Gilliam said, remembering the headlines.

In the *Tampa Tribune* on January 19, 1996, the activists' lawyer, Henry Valenzuela, labeled the action "an abuse of power" and added, "It's the ultimate act of big government power over the citizenry. I don't think the government should be suing citizens when they object to something."

While Gilliam and her lawyer enjoyed and inflamed the bad press Pinellas was getting, Eileen Hart was concerned about COLA. She was afraid that by personally suing several members, Pinellas would be successful in scaring people into silence. "People couldn't afford to be sued," she said. "After the original discovery, when COLA was helping the District, we were in debt. We couldn't afford to fight a legal battle personally. Nobody could."

Because people were afraid, they quit supporting COLA. They didn't want to talk to reporters; they didn't want to participate. Although it may not have technically been a SLAPP suit, it had the same effect of discouraging public participation.

The meeting where the Pinellas commissioners made their decision was a clear memory for Steve Seibert. "It was an afternoon board meeting. It was dark outside. Ed de la Parte had briefed me beforehand, and my stomach turned. I should have listened to it. Before our board meeting, I'd had some time to think about it. He told me, and he told the board, 'We need some defendants in order to get a declaratory judgment.' I asked the right questions. 'We're not seeking damages, and we're not trying to hurt anybody?' 'No. No,' he told us. I voted for it. Bob Stewart, to his credit, did not. This is a good lesson: don't exercise the awesome power of government without thinking it through," Seibert said.

Both Ed de la Parte and Doug Manson insisted that the Pinellas action was not a SLAPP suit. Commissioner Seibert, the lawyer, knew the difference, too. "We didn't try to stop them from speaking; we tried to get a summary judgment to determine whether we had liability," de la Parte explained. He meant liability for wellfield damage and damage to private property near the wellfields—damage that Pinellas, St. Petersburg, and the Authority had denied. Still, the move, while legally explainable, was publicly indefensible—a public relations disaster.

"They asked us if we wanted to participate," Mayor Fischer said. "We didn't want any part of it. I wouldn't sue citizens; I wouldn't do that. Pinellas was going down a very combative path." He trailed off, thinking, perhaps, of the path not taken.

"It turned out to be a huge political mistake," said Manson. "By itself, it wasn't enough to topple Ed's influence with the commission, but on top of everything else, and the number of people gunning for him, I think it marked the beginning of the end of his dominance over water policy in Pinellas County."

Still, de la Parte maintained it was not a mistake. "We did not put the activists at risk," he insisted. "We asked a judge to tell us if we were liable for the damage. If he said we were, we would have made the situation better for them. If he said we were not liable, then that would have barred claims against us."

In fact, the commission made the move more than once. It sued the citizens, then when it lost that case, it appealed. From a legal standpoint, it had to, or it would have been subject to considerable liability. It needed the courts to hold it accountable for environmental damage or exonerate it. It was trapped by its own tactic and had to appeal. Eventually, Pinellas County settled with the activists. "They settled because the atmosphere changed," de la Parte said of the Pinellas Commission. As part of the reorganization of the Authority, de la Parte added a provision that had the Authority paying 56 percent of any judgment against Pinellas in the case.

"It was a classic case of listening too much to your staff," Hillsborough commissioner Turanchik observed. "The Pinellas Commission took a lot of heat."

But the lawsuit was successful in splitting the activists. Several settled right away, fearful of Pinellas County's power and money or just tired of the hassle. Three of the original eight said they were staying until the end, whenever that was. One, Judy Williams, said she would continue until she got an apology.

She joined the water wars in 1994. She came to a Governing Board meeting after Steve Monsees's presentation but before the emergency order. At her first presentation, she was wearing cutoffs, a T-shirt, and Keds without socks. She had a two-year-old on her hip. He was screaming so loud she had to leave the boardroom. She missed her chance to speak the first time her name was read from her blue card. Someone went in the hallway to get her. She came in, waited until the end, and walked to the podium. She apologized for the child who still hadn't quite settled down, and on her tiptoes she said quietly into the microphone, "My lake is dying. I don't ask you to do something for me but for him." She nodded at the boy in her arms. "It's my home, but it's his place, and you can do something to stop what's happening. We're counting on you." Three minutes go by quickly when you're pouring your life, your heart, into a microphone.

In 1999, Judy finally got her published apology as part of the settlement agreement. Pinellas paid $341,600 to the activists. Together Gaye Townsend, Meredith Wester, and Judy had a separate agreement to share the money. Meredith took part of hers and bought a water fountain for her front yard. It's a permanent memorial to a failed water policy and bad political judgment. The *Tampa Tribune* ran a story on May 13, 1999, and the *St. Petersburg Times* followed on May 15.

"Bottom line, it was ill advised," Seibert says. "It was ill advised by our lawyer, it was ill advised by our board, and it is a vote that I am not proud of. But we did it."

Water Problems Plus

Just when pundits thought Pinellas County's mistakes couldn't get any worse, they did. Angered over its latest legal maneuver being called a SLAPP suit when it technically was not, it decided it needed some public relations support. Determined to get its story out its way, it called for proposals to determine how best to proceed. It was not the first time that Pinellas had thought about public relations. Just three years earlier, in 1992, in anticipation of the next review of their permits, Pinellas County leaders developed a request for proposals (RFP) to have a public relations firm do several specific things. The situation analysis in the RFP includes this background information:

> Pinellas County comes under the jurisdiction of the Southwest Florida Water Management District. SWFWMD is in the process of changing from a resource agency into a regulatory agency. At times, these rules and regulations seem arbitrary and unnecessary. They do not take into account other, more environmental and conservation alternatives . . . In the future, should Pinellas County challenge the SWFWMD rules and regulations that place unnecessary burdens on the Pinellas County residents, charities, and businesses without seeking sensible alternative solutions, the County wants to have a solid reputation as an excellent water management organization and avoid being portrayed as an anti-conservation and anti-environmental entity.

There is no paper trail to suggest that the District warned St. Petersburg, the Authority, or Pinellas that changes to their permits were coming. But the RFP indicates that as early as 1992 someone knew, or at least suspected, that changes were coming. And someone wanted to do something about it before the review of the Pinellas permits took place. Someone thought that public opinion could be central to success in any future confrontation. One of the stated goals in the RFP included "To build a strong, grassroots coalition of concerned Pinellas County citizens and organizations willing to lobby and/or confront SWFWMD regarding arbitrary or unnecessary rules and regulations."

Was Pinellas gearing up for conflicts to come? The RFP for public relations aggressively aimed at the District, the rehiring of Pick Talley (in 1992), and the content of the RFP all suggest that someone was anticipating conflict.

In 1995, in the middle of the water wars, a copy of that old RFP showed up in my mail. Helpers. Unknown. Providing information and insight. I can only wonder what would have happened if Pinellas has successfully implemented the goals of its RFP.

Despite its failed earlier attempts at public relations, by 1996 Pinellas sought public support for its water policies and was willing to pay for it. In a Pinellas Commission work session discussing the commission's alternatives, Pick Talley introduced Cedar Hames of the advertising agency Earle Palmer Brown as they started their discussion. Cedar began by reminding the Pinellas commissioners that their own public-opinion research, completed in 1995, indicated that people really believed that there's a water problem, that seawater desalination was a good choice, and that reclaimed water was a possible alternative. Hames took the board members through potential budgets and programs, with a proposed cost of $1,350,000.

Commission chair Sallie Parks asked about timing given the legislative session, scheduled to start March 1.

"Yes, I think we need to get out and tell the county's story," Hames said. "Swiftmud has done a very good job of promoting the low water levels and the fact that it's hurting the environment. The [proposed] Pinellas legislation comes up toward the end of the session, and so it would be appropriate to communicate that to the public, make them aware that there's a lot of gray area as to whether it's pumping or development causing a problem."

"I'm very reluctant to do this," Commissioner Seibert said, still stinging over the SLAPP-suit headlines and editorials.

"I think it's very regrettable that this issue comes to us in this form at this time because of the action we just took or have indicated we're about to take with regard to the Coalition of Lake Associations," said Commissioner Stewart. "If I thought that the amount of dollars that we would spend on a public relations campaign would have a positive effect on either of the actions we've just discussed, I'd be asking for more than $1.3 million to spend on it," he continued. "I guess the only thing that I'm saying is, we do need to have an orchestrated educational program for the Pinellas water consumers and for the legislators . . . but I'm hard-pressed to go with [it] now."

Pick Talley responded quickly, his soft, southern accent barely concealing his anxiety. "The urgency is . . . when we go before forums in an administrative-hearing process and any of these other legal processes, there's a presumption that Pinellas County caused the problem. We don't think that's a fair one. We don't think it's been reported with balance." Clearly getting to the heart of the policy differences, Talley explained, "If we can't rely on the groundwater, if we have to go to alternative sources, then whether it's these projects or others, we're going to be doubling our water rates over the next twenty years. Those decisions are being talked about now, and it's our view that if you wanted to have an impact on those, we've got to move."

Then, acknowledging the public's strength on the issue, Talley concluded, "Even if our lobbying program works, we need to create the idea that this is a debatable issue. We may not convince everyone that water's not scarce . . . but at least they need to hear the other side, and right now, the District has the whole field."

"It's a debatable issue." That's what he told Senator Latvala.

"What does a program like this do to your legal argument?" Commissioner Rainey asked Ed de la Parte.

"I think it helps us," he said. "Like Pick, I think there's a perception out there that it's not even a debatable issue, and I think that you need to at least bring public opinion back to neutral. I would hate to bring these matters before a Pinellas jury only to find out they've bought into the District's propaganda. We've allowed two years to go by where the District had a free shot. We need to begin the process to reclaim that ground. I would recommend it be sooner rather than later."

"Swiftmud's efforts are very professional," Commissioner Rainey said. "It's going to be hard to catch up. Very hard."

"What better time to start the program than when we've had very abundant rainfall and a lot of the lakes in the area are very full?" said de la Parte, recognizing a strategic opportunity. "If you wait and we go into a droughtier situation, then you lose the effect . . . I think it's something you need to take into consideration."

"What will it cost us if desal goes through?" Commissioner Tyndall asked. "Swiftmud board members have threatened to cut us back by as much as 50 percent, regardless of the science. Swiftmud has been pounding on us unmercifully. Honey Rand deserves an Emmy for her 'one flush.' Totally incorrect, but it rang a bell through everybody. I talked with Senator Safley a few months ago, and he says when he tried to speak [in Tallahassee], they said, 'Slurp, here comes Pinellas County.' You know, we're slurping all the water . . . we're getting killed PR-wise. It's been heavily against us. Even our own citizens."

"So if you don't adopt this," Rainey added, "there's going to be no rebuttal, because Swiftmud is running rampant, and they've spent an awful lot of time with the editorial staff of both newspapers. West Coast [the Authority] hasn't done anything . . . as far as the public goes." He said to Cedar Hames, "I think your presentation is weak. I would like it to be a lot tougher . . . but I think it has to be done."

Fred Marquis, the county administrator, added, "Your Tallahassee consultants have also reviewed this and strongly urged you to take advantage of this opportunity." Marquis was the longest-serving county administrator in Florida. He was among the few who kept their jobs for more than twenty years.

"I can count," Seibert said, recognizing that this vote would likely pass. "I'm still uncomfortable with this and think there are better ways to get the point across. But it appears the wisdom of the board is otherwise."

In his excitement that the big-dollar project was going to pass, even Cedar Hames had to hesitate when describing the situation near the wellfields. "You'll possibly see our work in announcing the court case coming up," he said, perhaps the realization dawning on him that a lot of money would flow to his firm. Maybe he was excited about the challenge. He spoke a bit faster. "Another is the picture in the paper this week, with the dock going out into the landbound area and everyone standing there in a group. It might be interesting to show a similar photo of a lake that isn't affected by the. . ." He hesitated, trying to pick the right word, failed to find one, and moved on. "And then put those two photos side by side." He just couldn't force out the words "wellfield pumping" in front of this group. In the end they voted to spend $800,000 to explain that there is no environmental damage, no water shortage, just a shortage of facts. But his hesitation suggests that he, too, was convinced that wellfield pumping was a significant factor in the environmental damage. The District's message had gotten through.

The District learned about the planned PR campaign through the newspaper when the *St. Petersburg Times* reported on the commission meeting. On January 18, 1996, headlines heralded the commission's latest effort: *"Ads will pump the water word out. The County will spend $800,000 on newspaper, radio and television ads telling people that the Bay area doesn't have a water shortage problem."* Ed de la Parte added that an effective ad campaign could help sway potential jurors in any water-related lawsuits. Given the commission's vote to file suit against the water activists, COLA, and others, the idea of a PR campaign was tainted. Many more people were appalled at the content when it launched.

A week after the commission's decision to spend $800,000 on a public relations program, Pinellas commissioner Stewart noted during a commission meeting, "We've taken a pretty significant bath in the newspaper for some of our recent actions in relation to water. The suggestion to send a letter to Swiftmud inviting a nonpartisan, academically oriented group to seek the truth—that's a very significant step. I saw a ray of sunshine there. And I think that our educational program is another attempt at trying to broaden the base of common knowledge. So I see a number of things that this board is taking the initiative on that place us in the forefront of trying to reach out rather than trying to retrench and stick our heads into the sand and being totally parochial about the issue. I'm totally supportive of the direction we're heading."

After his comments, a Pinellas resident took the microphone and concluded, "I've spent forty years as a Floridian, thirty-six of those in Pinellas. I'm not totally in the dark about Swiftmud and the Authority, but I'm concerned about this basic attitude here in Pinellas that we don't have a problem. My comment is, I have a great deal of concern with this decision for an $800,000 expenditure for public relations to tell the Tampa Bay area that we don't have a drinking-water problem. Any questions?"

Pinellas County's water-supply problems and public relations problems were bad, but it hadn't hit bottom yet. The permits that were at issue weren't even its; they were St. Petersburg's. Its permits were next.

Were Pinellas County commissioners following their collective conscience, or were they stuck in a quagmire of their own making? Were they following the lead of Chuck Rainey, or had they lost their ability to anticipate consequences? Was their behavior ethical? Not in the basic sense of the word. But in a broader sense, was it ethical because they were trying to protect their constituents from additional expense, even if those constituents had weighed the alternatives and expressed willingness to pay, something the District asked in its public-opinion research?

Chapter 14
In the Future, Nothing Will Be the Same

Steve Seibert laughed when he thought about it. "Chuck Rainey was such a tough guy. It had to have been hard for him to go through the visioning exercises and all the touchy-feely things we were doing at the board retreats."

For all the public water disputes, Rainey was losing at the Authority and with the District, and by 1996 Steve Seibert was openly challenging him on water issues at Pinellas Commission meetings and retreats.

"We respected him immensely," Seibert admitted. "He was the elder statesman, and he had always delivered on water for Pinellas County. But there were too many questions, too much was going wrong, and people were asking me, 'Aren't you watching the store?' It was a tough time."

"There was this change going on," Seibert acknowledged. "In our closed-door sessions, and we had a lot of them, there was this real tussle, and it was generally Chuck and me."

Sunshine Laws allow elected bodies to meet privately in executive session for matters having to do with legal strategy. The sessions were taped, and when litigation was complete, the materials were made public.

"There was an ideological shift occurring, and Chuck said, 'Look. If we're going to fight this battle, let's fight to win. If you're going to compromise and mediate it, do that. But do it with all your energy. Don't try to do both.'"

The comment hit Seibert hard. "He was right. I was trying to fight the good fight and put out the olive branch at the same time. It wasn't effective, and we were getting killed in the process. And [the District] was capitalizing on our mistakes. For a time, everything we did was wrong."

If being challenged on water issues wasn't enough, as part of the Authority board Rainey had to help select the new water sources. The Authority was holding its own visioning sessions. He didn't like it—the alternatives or the sessions.

"There's plenty of groundwater," Rainey complained. "We didn't need desalination. We didn't need the wastewater project. People have told me to eat shit before, but they'd never told me to drink it."

He was serious even though the line was always delivered with a smile. He repeated it in meetings, with the *St. Petersburg Times,* and to anyone who wanted to talk about water. By this time, the Authority general manager Maxwell was convinced that water supplies needed to be diversified. But getting the Authority member governments to agree on those projects—and agree to pay for them—was a different matter altogether.

Rainey was not happy about making choices among projects he didn't want, but every member had to do it.

"We gave each board member a certain number of red stickers or magnet pieces," Maxwell recalled. "Commissioner Rainey didn't want to do it, but everyone was participating, so what choice did he have? When it was his turn to put the dots next to the projects, he took the yellow dot, turned, and faced the group, raising his middle finger while showing the dot, and turned and put it on the board. He did that for all the projects," Maxwell said, laughing. "It was typical." Each "salute" was aimed at him.

Rainey didn't remember the day, but he wouldn't deny it either, saying instead, "It could have happened."

"He called me Commissioner Tourniquet," Ed Turanchik said, laughing. "He's a funny guy."

Back in the 1970s and '80s, Hillsborough County had given away its rights to develop water sources independently. Now Hillsborough officials wanted them back. Hillsborough County even offered to support disbanding the Authority if it came to that. Coming from Turanchik, the chair, it was a threat.

"Chuck didn't like it at all," Turanchik said.

Turanchik believed the Authority was established to serve Pinellas County's interests. The structure of it made sense no other way. But the days of rolling over Hillsborough and Pasco Counties to benefit Pinellas and St. Petersburg were gone. The times and the people had changed. Turanchik confronted Rainey.

"Every new source that we try to develop that isn't simply a wellfield permit or pumping increase, Pinellas objects to it," Turanchik argued. "There's always a reason to object. Always a reason to stop the project."

Determined to make his point, Turanchik went to Tampa's Mayor Greco. He got Tampa's agreement to disband the Authority. The vote would be 3–2. Rainey would lose. But it never came to that. The Authority let the matter drop. It was revisited later—during reorganization negotiations when the possibility of dissolution was always there if an agreement couldn't be reached. Tampa Bay needed new water supplies.

"From Hillsborough's standpoint, freezing the Authority made sense—especially if we developed our own water supplies," Turanchik said. "But disbanding or freezing the Authority made no sense from an environmental standpoint, and that's what we were trying to change." In other words, when change came, it wasn't sufficient to take care of the future; they had to find a way to repair the past. The Authority needed enough water for a secure future, but it needed more water to reduce pumping from its present water-supply sources.

"It was hard," Turanchik admitted. "Maxwell worked for me, and yet my local government was suing the Authority. The whole thing was ludicrous."

Chuck Rainey had been the power behind the politics of water in Tampa Bay for nearly thirty years. When Rainey was first elected to office, Commissioner Seibert was in junior high school. Of all the involved parties, he was the only person who had been an elected official all that time, although Tampa mayor Dick Greco had served before, and water lawyer

Ed de la Parte had been a player since the mid-1970s. In October 1996, Rainey told reporters he had health problems, and he was leaving the commission. In truth, things were changing around him. "I couldn't make a difference anymore," he said with a shrug. "What's the point of staying?"

In leaving, he sent a note to Steve Seibert that said, in part, "You always held your position in arguing your point of view. In our world, that's to be most respected."

"That is among my most treasured possessions," Seibert said. "Man, from Chuck Rainey, that really meant something."

As Rainey announced his retirement two years into a four-year term, Steve Seibert was chair of the Pinellas Commission. He appointed himself to the Authority. Along with Ed Collins from Pasco, Ed Turanchik from Hillsborough, and Dave Fischer from St. Petersburg, everyone was hopeful that the internal fights at the Authority could be resolved and new water-supply projects could move forward. There was a lot of work to be done, but at least these four people gave every indication that progress was possible and that together they could make water work.

"We needed a new cast of characters," Turanchik said. Still, he was stunned by Rainey's departure, as was everyone else. It was the end of an era. "Rainey's departure made a huge difference," Mayor Fischer said. "It was a big switch from the way Rainey did business to the way Steve Seibert did. Seibert is the kind of guy that wants to work things out. The tenor of the board changed overnight. Chuck left, then Commissioner Tyndall left, and boom—it was a different commission," according to Mayor Fischer. Commissioner Tyndall resigned after being caught in a law enforcement sting. Governor Chiles appointed a replacement.

"At some point, we just realized that it was on the legislative agenda and that we had no support," Steve Seibert said.

"It was one of those windows when people have respect for one another," said Governing Board member Roy Harrell. "It was a window of time, an opportunity, and they did some really good things." Harrell and Rainey had been friends for many years. Harrell disagreed with Rainey's water policies, but they had worked together on other community issues. While the District was preparing and issuing the emergency order in 1994, Roy was helping the dreaded Commissioner Rainey in his campaign for reelection. It was hard to believe that Rainey wouldn't be part of water policy in the future, but everything was changing.

Harrell became the chairman of the Governing Board in the summer of 1996. It was a long-awaited goal and his turn to leave footprints in the sand, something he talked about frequently. As a lawyer, Harrell had few employees—a paralegal, an assistant, and a secretary. Coordinating that small group and coordinating the activities of a 750-person agency are very different. With 3 people, you can take the pulse of employees' thoughts and objections. You cannot do that with 750 people. As Executive Director Hubbell had said years before, "You can't turn it on a dime, or even in a week."

"It wasn't so much that Roy would turn on a dime," Assistant Executive Director Mark Farrell observed. "I mean some of Roy's ideas were fabulously great, and others were not. Roy always had a great sense of politics, without question. But once he locked on to an idea, he was a bulldog. As staff, we had to get to him early before he formed a strong opinion because it was hard to change his mind. Sometimes we did, often we didn't," Farrell said in that sometimes-the-bear-eats-you way. "Roy knows how to approach a problem the right way. On the downside, he was too trusting. He believed that when it came right down to it, people would do the right thing. They took advantage of him."

But Harrell had a plan that he hadn't shared with anyone. Thinking back on it, St. Petersburg mayor Fischer thinks Harrell sensed the opportunity to move the agenda forward. "When the change in the Authority board came, Roy saw the opening," Mayor Fischer observed. "I can just hear him thinking, *'This group is going to change direction; if we partner with them and help them put their plan in place, we can drop the lawsuit and develop some water.'*"

What Harrell also saw was a board with ideas that might not include the District's desire to develop alternative water supplies. "They had some ideas, but they couldn't seem to get them moving," Harrell said. "There was the master water plan, the Authority's projected agenda for water development, but they still had internal problems and problems at their respective commissions and councils."

Harrell strongly believed that desalination had to be part of the long-term regional solution, and he had an idea about how to help achieve both.

Everyone shared weariness with the water wars, the siege mentality, the sniping in the media, and the political hardball. The public was tired, District staff were tired, and elected officials were tired. The only people that gave no indication of letting up were Pick Talley and Ed de la Parte. Still, the sides seemed stuck, unable to gain closure in their ongoing negotiations. Both St. Petersburg and the District wanted to make a deal to end the litigation over permits but couldn't do it in isolation. Any agreement had to be returned to the respective boards for approval. And there was always Pinellas County, adding objections and obstacles through its lawyer.

"Everyone involved in this thought they were doing what they could to serve the people," Steve Seibert observed.

"About three weeks before trial, we had a meeting that included Mike Davis, St. Petersburg's city attorney; Dan Fernandez, the outside counsel; and the mayor," said Doug Manson. District Executive Director Peter Hubbell, Governing Board member Joe Davis, and attorney Jim Robinson visited to give it one last shot to work out the problems. "The District had proposed to phase down pumping over time, like five years."

"They said that if we didn't have replacement supplies at the targeted levels, we could continue pumping, not at the permitted quantity, just at the level we'd been using," Manson said.

St. Petersburg's team was concerned about what would happen if they couldn't meet the deadlines. They asked the District to put the offer in writing. "We can't do that," Joe Davis said. "The Governing Board will never agree to it."

"You'll have to trust us," Hubbell told them.

But, of course, they could not just trust the District. There was far too much at stake. St. Petersburg officials needed the agreement in writing, and the District wouldn't, couldn't, do that.

"I couldn't take that kind of risk," Mayor Fischer told Davis sadly. "I have a responsibility."

"At worst you would just keep pumping," Davis told him. "You know we can't turn off the pumps."

"I'm a public servant. I can't agree to something I think we'll violate in the future," the mayor said. And the meeting was over.

"The mayor went in there to make a deal, to settle the litigation," Manson said. "The District side blamed Dan Fernandez for not pushing the mayor to do it, but how could he? He could not agree to something he didn't know he could do. He just couldn't."

They just couldn't see a way out.

"The trust-me argument was lame," Doug Manson said. "I felt like if we had just a little more time, we could have reached some agreement. If they had taken it to the board, they could have had a 5–4 vote or something."

But a 5–4 vote would have publicly indicated the Governing Board was conflicted. The District had gotten this far by showing a committed, united front. The District Board believed it was right and that it would prevail. Besides, the toughest issue wasn't really St. Petersburg's permits. The District was confident that eventually a deal of some kind would be worked out with St. Petersburg. The real challenge was the Authority and Pinellas County.

At issue was the cumulative impact of the entire water-supply system and the failure to build sufficient new water supplies. The St. Petersburg permits, everyone knew, were a test case. Ed de la Parte was testing his arguments on behalf of Pinellas County. The Pinellas permits were next.

"To some extent, I understood," Manson said. "Everyone knew we would try our best to meet any cutbacks, but what about the Authority? We were part of the regional system, and if Pinellas blocked it again, we'd be in the same position five years from now with no new water supplies. The mayor felt that cutbacks were inevitable. St. Petersburg was beyond denying there was a problem. The real question was, what are we going to do about it?"

There was constant pressure on St. Petersburg from Pinellas and its representatives to hang tough. "You're not going to settle" came as both a question and a threat. "They couldn't really threaten the city," Manson observed, "but they could keep up the pressure." They did.

Divining the Waters

After fits and starts, delays and negotiations, the four wellfields cases got underway. It was September 1996. The District felt that it had a strong case and was ready to face whomever they needed to face. Doug Manson represented St. Petersburg, Don Conn represented the Authority, and Ed de la Parte represented Pinellas County. Pasco, Hillsborough, and the District had inside and outside counsels. Ed de la Parte always stuck in the District's craw. Every time he opposed the District, he beat them.

"It became a kind of mental thing for the District," Manson speculated. "They couldn't beat de la Parte, so they demonized him. What's a demon? Something you fear. When the lawyers couldn't get the job done, there was a concerted effort to separate him from his client."

District staff didn't like de la Parte or what they considered his slash-and-burn tactics. Some said it wasn't the District but his high fees, the suit against citizens, or his obstruction of regional cooperation. It all contributed to his public persona.

De la Parte and his supporters saw it differently. "I was a lightning rod because they knew they couldn't beat us," de la Parte said. "The only way for them to win was through public relations. And what better way for you to win the public relations war than to demonize? And so, they—*you*—did that to me," he said. "I know where the skeletons are, and I do my homework. I've never lost a water-use case or a rule challenge regarding water. And the District knows that."

This was not delivered as a boast. It was a matter of fact. At least it was a fact until September 2000, when the District prevailed in an appeal over new water-management rules in the Southern Water Use Caution Area—rules developed and challenged during the water wars but in an entirely different geographic area. It was the first time the District beat de la Parte. It was a total and (for District staff) sweet victory.

Manson observed, "The first reason you go after someone is that they are an excellent opponent. If you can't beat them in the courtroom, you go after them in the press . . . after fees or whatever." That frequently happened to de la Parte over the years.

As the case was underway, Ed de la Parte was examining one of his witnesses, Shirley Denton from (then) Biological Research Associates. As part of the normal discovery process, documents were previously provided that supported her testimony. Among those documents was a memo from de la Parte to a consultant (dated years earlier) that read, "When you go, you will find. . ." The document was open to interpretation. Either de la Parte was telling the consultant what to find, or he meant it as "I believe you'll see. . ." Could a meticulous lawyer like de la Parte have accidentally written the memo with that kind of ambiguity? Would he have told the consultants what to find? The District believed the latter. So did a lot of other people. On August 13, 1996, the *St. Petersburg Times* carried a story on the document.

"It was a low point for the District," de la Parte said. "You were sitting in the audience with the reporter, to get her to observe these questions, and you handed a bunch of documents to her. The District was going to do anything to win because they were not going to win in the courtroom."

The staff told me to do what I could to get him off his game. It was theater. There was nothing much in those documents, but it unsettled de la Parte. It worked. I'm not going to make excuses for handing the documents to the reporter, but I do have to say that my presence was more for show than anything else. I didn't need to be in the room to give papers to a reporter. I was there because I was distracting to Pinellas County's lawyer. I was there to get him off script. Did it bother me at the time? No. Despite Ed's later claims that District lawyers said they were ashamed to be part of the organization because of that action—that was not the talk inside the agency. Inside, people were glad that the public would see that Pinellas consultants would take direction and did. To the District, they finally had proof of what they always believed. They called some of the scientists and technicians working for the Authority and Pinellas County "biostitutes." They believed that the consultants did what they were told, and they believed that de la Parte was calling the shots. De la Parte and Chuck Rainey played hardball. For the first time, the District was in the same game.

***

Two years before, de la Parte had charged Governing Board members with Sunshine Law violations. He claimed he had a moral duty as an officer of the court to report such violations. Maybe that's true, but few people believed it was a major factor in pressing his concerns. If he could have proved they did something wrong, any votes they took would be voided. And board members were embarrassed as his accusations made the news.

De la Parte sees a clear distinction between my providing a memo to a reporter and his pursuit of Governing Board members. But to the board members, and the people who cared about them, there was no reason to give him the benefit of the doubt. *St. Petersburg Times* reporter Jean Heller is an experienced hand in regional hydropolitics and was sufficiently convinced to write about it.

What was it about Ed de la Parte that made it so believable that he would order environmental consultants to produce specific findings? Some of it surely had to do with the fact that environmental damage on and off the wellfields was abundantly clear yet Pinellas continued to deny it. Many years ago, I read that a smear doesn't have to be true; it has to be believable. This was.

"I remember the morning that article ran," said Doug Manson. "I was in Ed's office, and we were all outraged. When I walked in, he was sitting at his computer, typing. He just kept going while I ranted and railed about it. Finally, he turned to me, and he was as upset as I have ever seen him, and said, 'What do you expect me to do? Anything I do is only going to make it worse.' He was right. There was no response."

Then, Ed de la Pate was embarrassed a second time.

A short time later one of the paralegals working for de la Parte asked a custodian to let her into the District's temporary offices, set up for the trial. The two legal teams had shared documents, and the District had some that were needed. She was told to go get them. The custodian, known as Elvis, let her in, then told the District's legal staff.

The incident made the news on September 21, 1996.

"That whole thing was much ado about nothing," Manson said, clearly disgusted with the subject. "We were in each other's offices all the time. We were told she went there looking for an exhibit that the District borrowed from us. At that point, excuses for nonperformance were just not accepted. If you were told to get a file, you did it without complaint. But I can tell you that Ed de la Parte did not instruct her to go there. It's not his style."

Despite the headlines, the trial continued. As with virtually everything related to the Tampa Bay region's water-supply issues, each side tried a different case. The District, Hillsborough, and Pasco tried a case of science and environmental impacts, while the Authority, Pinellas, and St. Petersburg tried a case that centered on impacts that had been created long ago—and asserted that any impacts had long since stabilized. Besides, the District, through the continued reissuance of the permits, had allowed the impacts, they claimed.

"The whole case was not about whether the wellfields caused impacts," de la Parte claimed. "The District tried that case, but the Authority didn't. The whole case from our perspective was, will the continuation of those withdrawals create impacts which are significantly different from what has occurred historically? That was important because the District knew those impacts would take place, and they determined the impacts were acceptable." For emphasis de la Parte pointed out that even Mark Stewart, a University of South Florida professor and consultant to the District, testified under cross-examination that the environment had reached a dynamic equilibrium fifteen years ago and the situation wasn't going to get any worse. Stewart had been the District's top expert.

"Some people thought that we were saying it's OK to impact the wetlands," Manson said. "Our experts didn't say that, and they didn't say we had no impact. All I wanted to do was show the hearing officer that there is a dynamic, vital ecosystem out on the wellfields in comparison to what's on the other sides, which is mostly development . . . and that the rate we were pumping wasn't going to increase any problems."

His biggest fear, Manson said, was that the District would be successful in demonstrating the wetlands had been annihilated and that the hearing officer would say it couldn't go on.

"We tried to paint as accurate a picture as we could," Manson said. He recalls an exchange between District consulting attorney John Wilcox and his own environmental consultant, Jim Bays. "John Wilcox read the standard for an adverse impact to Jim Bays and said, 'Well, Mr. Bays, does the District take into account the total function of ecosystems, wetlands, and uplands?' and Bays shook his head and said, 'Sadly, no, it doesn't.'" According to Manson, the exchange indicated that the District was aware that there were impacts when it issued the permit. The standard of adverse impact had already been met when the permit was issued, so the District had in effect approved that level of impact.

But the District's biggest mistake, according to Manson, was denying public-supply permits. He thinks that had the District stayed with the stepped-down quantities over time, it might have given the judge an alternative. Then again, the burden of proof would have been on the District, and de la Parte had already promised that Pinellas would challenge it—and win. Like so many junctures in this conflict, there is no way to know where the path not taken might have led. "Denying the permits helped me paint the District as draconian, as unreasonable," Manson said.

It was the pressure of litigation that drove the nastiness, and it was the same pressure that was driving an alternative solution. "It was the litigation that made an accord possible," Hillsborough commissioner Turanchik insisted. "If the District, Hillsborough County, and Pasco hadn't teamed up, if we hadn't carried a big stick, we could never have forced a change." On the one hand, local governments needed to believe the District would take this to the end; on the other, the District needed the public to be absolutely convinced no one would ever turn off the water supply.

It was a fine line that no one dared cross.

But lots of other lines were crossed in the process. In the effort to win, participants painted each other's actions in the worst possible light. People said I destroyed Ed de la Parte's invincible facade. At the time it didn't bother me, but years later when I was asked to do it again, I said no. Never again, not for de la Parte, not for anyone. Win on the merits or lose; let people get the reputation they deserve based on their actions rather than personal attack.

Chapter 15
Dividing the Waters

Despite the hysterics promoted by Pinellas County, no one believed the District would turn off St. Petersburg's water or anyone else's for that matter. It can't be done. Water systems are demand based. When the pumps are turned on, the water flows, if there is water at the source. But, according to one perspective, the hearing officer did believe it, and the District didn't leave him any alternative. The goal of denying the permits was to exert the District's authority to change things. After all, the point of having permits periodically reviewed was to ensure that they did what they were supposed to—provide water for a specific purpose, for a specific time, under specific conditions. The fact that permits must undergo periodic review suggests the water-management system is there to identify any unintended consequences. If the District has no ability to modify a permit, what is the point of the review? For that matter, what is the point of permitting?

When the hearing officer's decision was rendered on the four-wellfield case in 1997, it was almost anticlimactic. Reorganization of the Authority was being discussed, as was the District's offer to financially support new-water-supply development. Each side had tried different cases, each got the same decision from the judge, and each read that answer differently. The District, Hillsborough, and Pasco said the science proved the wellfields were causing widespread environmental damage. The Authority, St. Petersburg, and Pinellas declared victory because the judge told the District to issue the permits at present quantities. What the judge said was that environmental degradation had occurred because of wellfield pumping, but it was no worse than the last time the permits were approved. So he ordered the District to renew the permits at existing quantities. But he pointed out there was insufficient evidence to support any future increase in permitted quantities.

"The thing that persuaded the judge was that the impacts were not that much different than were predicted when the permits were originally issued," de la Parte explained. "When the time came, Ted Rochow, God bless 'im, would have testified that he wrote memos to his supervisors saying he thought there would be these kinds of problems. At the time, the District had decided that these were not significant adverse impacts. At the time, twenty years ago, the District wanted these projects." The District wetlands ecologist, Dr. Ted Rochow, had spent more than twenty-five years monitoring and documenting changes in wetland ecology on or near the wellfields.

"The wellfields were, in fact, pumping less than they had historically," de la Parte said. "If anything, the impacts improved when the historic pumping decreased. At the very least, they were no worse. That's why the judge capped the permits at their present limits." The judge said an increase of even two or three million gallons per day might exacerbate the impacts.

"After these permits," de la Parte went on, "we would have tried the Cross Bar and the Cypress Creek cases, and the District would have lost those, too. People who wrote memos to the files are no longer at the District. Those people would have come to testify."

Imagine all those years later coming back to testify for Ed de la Parte that, yes, what you predicted had happened. Now wetlands are dead, lakes are gone, and the environment has been irreversibly (in some places) damaged. Imagine what it would have been like for technical experts who were trying to protect the environment, to stop or curtail the pumping with those long-ago memos.

"That's where Mark and Pete made their mistake," de la Parte concluded. "They thought these impacts were something very clearly communicated to everybody back in the late 1980s, and they were wrong. No way. It wasn't communicated to the Authority."

But there is circumstantial evidence that the attempt to communicate was made. According to Gene Heath, the Authority's general manager from 1978 to 1990, the Authority had its own experts with their own opinions. Did they notice the environmental changes? Did they care? Why didn't Authority experts share their concerns? Or if they did, where did those internal memos go? John Parker recalls raising the issue of declining water tables around the wellfields with an Authority engineer at a technical meeting in the late 1980s.

"I was trying to tell her about the technical problems," John said. "I showed her the data and began to explain it . . . and she said, 'Now hush, John.'"

"Now hush, John," at a technical meeting? Was it personal or the result of arrogance from the Authority staff knowing that the water supplies would stay online regardless of District-staff concerns? That personal exchange was not part of any public record. But John Parker remembered it, and it was reflective of the two entities' relationship and missions.

But de la Parte was right on one critical point: there was no clear historical record nor public documents created by the District about its concerns. There were only the internal memos, which, when revealed, provided exactly what the authors didn't want—a reason to continue the pumping. Even in the December 1993 meeting between former Authority general manager Harold Aiken, his staff, and Pete Hubbell, concerns were communicated, and the Authority dismissed them with a "Yeah, yeah. Sure, sure" attitude. Now, years later, with more than $10 million in taxes expended in litigation, the legal decision was broad enough to let all sides declare victory—but no one really won.

The District didn't want to issue the permits, and by now Mayor Fischer didn't want them, anyway. He didn't want permits that a judge declared had damaged the environment—even though the judge said no new damage would result.

"I never had the feeling that the mayor was pleased with the decision," Manson said. "From a public-communications standpoint, we didn't win, but from a lawyer's point of view, we got our permit. We won. Still, the mayor was in a bad spot. He had the water he needed. He could have gone back to the city council and said, 'We got what we needed; we don't have to negotiate anything.'"

Even before anyone ever saw the judge's recommended order, the politics had shifted. St. Pete utility director Bill Johnson felt vindicated. He wanted to go forward and get his permit, while the mayor was convinced that a regional, political solution needed to be put in place.

"To do that, Mayor Fischer had to take on the city council," Manson said. "They made it very difficult for him to move forward on the reorganization of the Authority. But he did it."

"We could have said, 'Look, we got our three wellfields; we don't need anything else,'" the mayor said. "All of the growth was taking place in Hillsborough and Pasco Counties. We were not going to grow. But what if the District was able to cut us back? What if we lost some production capacity? Unless we found a solution, we'd be in trouble even after we'd done things right."

The problem weighed heavily on the mayor.

"There was substantial risk, substantial cost, substantial exposure," Turanchik said. "Bottom line, it doesn't matter who's right or who's wrong; the issue was how great is the risk and how much risk can you stand? What if you're wrong and then St. Petersburg must develop a desalination plant on its own? The risks helped create the opportunities for solutions."

"Know that your water rates could go up by as much as 40 percent over the next few years while you help pay for regional supplies . . . or risk they could go up by 140 percent if you get cut back and have to replace your own supplies," Turanchik said. "The math was simple; the stakes were incredibly high."

"We toyed with going it alone," Steve Seibert said of Pinellas County. "But every scenario required the wellfields to continue pumping at a certain production rate, and we couldn't be guaranteed that. There was a side of me that would have tried to go it alone—if I could have figured it out."

The risks were great, the pressure to solve the problem was mounting—and the players were about to change again.

A Sea Change for Mark and Pete

I was at the Tampa Service Office on a Friday afternoon when Mark called.

"You have to come back to Brooksville," he said.

"Oh man. It's Friday," I whined.

"You have to come back," he insisted.

"Can't we do this over the phone? I give really good phone."

"No. It won't take long. Head back."

"Bummer. It's Mexican night." I tried again.

"It won't take long," Mark repeated.

Clearly making no headway, I caved.

"OK. But this better be quick."

"It won't take long."

When I walked into Pete's office, I knew immediately that something was up. Normally, with the two of them, one was at Pete's table, the other on the couch or in Pete's chair, but as I walked in, they both turned from the table to look at me. There were no papers on the table. Whatever was coming didn't have any public records to go with it. This was bad. Really bad.

"It's Mexican night. What's up, boys?" I asked, acting much cooler than I was feeling.

"Sit down," one of them said.

"This better be important. I'm going to miss happy hour." I took a seat at the table. "Jeez, you guys look serious. What's up?"

Mark led, "We're leaving the District."

I hesitated a moment, thinking back to April Fool's Day. "I get it: when you least expect it, expect it. You made me drive all the way back for this? I will kill you." I stood up.

"Honey, we're leaving the District," Mark said, looking straight at me. He meant it.

I looked at Pete. "It's true," he said. "We're going." I could see Pete's face twitching. They meant it. It was written on Pete's face.

"You mean it," I said. It was a statement, not a question. They took turns talking. Blah blah blah. It didn't matter what they said; the District's future passed before my eyes. The enormity of this, the implications for the District, the leadership void. There was no evident replacement. It wasn't pretty.

"You're taking me with you?" I said, hopefully.

"We can't afford you."

"Does Roy know?" Roy loved these guys. He counted on them to keep him straight, on the right track, to back him up.

"We're going to tell him on Sunday."

"When are you going to tell everyone?"

"Monday."

"We knew that we couldn't make the announcement before the case rested," said Mark. He meant the permit case that had just concluded. "We didn't know if it could be used against our guys in court, so we waited. When Jim Robinson called in, I told him."

Monday seemed like a lifetime away. "OK, I told them I need to know what your plans are. If you are going, you need a great send-off. Is there anything I need to know about? Is there something making you leave?"

"No. No. It's just time," one said.

The District was between significant court cases; our public-opinion polls were as high as they had ever been. They wanted to leave while they were on top. I guess it was a good reason. I told them what they had to do to make the announcement, what I would do. We agreed to talk again on Sunday after they met with Roy. Poor Roy. Mark used to call me to his office and say, "Do that thing where you predict the future," and I'd give him a real scenario and tell him how it would play out. That afternoon, I predicted the District's future. I told both what would happen to the agency and how it would happen. I totally missed the speed at which it would happen. I thought it would take longer to unravel. It started well before they left. I got in the car to drive home. I thought about the District and the people who had vested their trust in the notion that Mark and Pete were going to change things. The job wasn't finished, and decapitating the organization was about the worst thing that could happen. We were going to face a leadership void at a critical time.

***

They took Roy to a restaurant featuring a beauty competition. Girls in bikinis. Roy was blessedly distracted. He knew something was going on because they wouldn't tell him why they needed to see him on a Sunday. "It was one thing for us to tell Roy that one of us was leaving," Pete said. "But two of us at once . . . and Roy had been a friend all these years."

"He asked, 'What's so important? You guys are leaving?'" Mark remembered, eyes wide open, stunned. "We kind of choked, but then we told him. We had decided that Pete would lead the discussion."

"When I realized to my horror what they were saying. . . It was going to be terrible." Roy said. "I had just started the Partnership Plan for the new water supplies funded by the District], and I knew I needed to try a new direction . . . I knew it was going to be hard without them."

"Roy asked us what we were going to tell people because it was going to look suspicious," Hubbell said. "Then he asked us to stay one more year, then six months. We finally settled on January." It was September 1996.

"We didn't want people to think we were leaving because of Roy," Farrell said. "We were doing it for ourselves."

When water activist Gilliam Clarke found out, she thought, "This is it for desalination." It was the water source of choice for the District and the activists. The Authority was resisting. "I figured things were going to change."

To this day, some people still think the executives were forced out. They can't understand why two people would leave while they were on top. There were so many special interests after them that when they announced they were leaving, people assumed someone had been successful at ousting them. There were political interests were after them, but that's not why they left, at least not directly. The four-wellfield case was resting with the hearing officer; discussions about reorganizing the Authority—to make it stronger and more centrally organized—were underway, and the District was in a bit of a lull. Since Republican Jeb Bush had lost the governor's race by a hair in the last election, few people thought he'd fail again. The rules had changed for District executive directors in the interim. Hubbell, although hired by the Governing Board, would have to endure Senate hearings and gubernatorial approval. Farrell likely wanted to return to the private sector, and Hubbell, reading the political tea leaves, couldn't see a clear future. With Farrell tugging on his sleeve, Hubbell made the decision to go—but not because they were forced out.

"That left a chance for the District," Doug said. "It was an opportunity for change, an opening."

"We changed enough of the faces to make it work," Steve Seibert said. "That included Mark and Pete." Everyone had carved out such strong positions for such a long time; the players just had to change, Seibert concluded.

Still, from late September 1996, when they announced their resignations, to January 1997, when they were due to leave, time seemed to drag even as it raced by. It was worse than being a lame duck. They were there and working, but it was hard for staff to connect. Before the senior-staff meeting on Monday, they told the executive secretaries. A river of tears flowed but not enough to fill the aquifer or to keep them at the District. Everyone was stunned.

That morning a consultant was scheduled to outline an internal communications evaluation undertaken by the District. It was bad timing in one respect because the agency was reeling. On the other hand, it was an excellent opportunity to get at real issues in a time of great uncertainty. The preliminary results of the study later played a crucial part in the District's future.

The Supply-Side Solution

Before their departure, Farrell and Hubbell were briefed on Roy's idea, the Partnership Plan. It was a way to make sure there was enough money available for alternative-water-supply development in Tampa Bay. Roy Harrell always maintained that there would be no resolution to the water wars and no cutbacks from the wellfields until local governments could be sure that replacement supplies were available to them. He called this a "supply-side resolution." His idea was to take the New Water Sources Initiative money, accumulated over several years and available, to buy new water supplies replacement quantities. In so doing, it would be impossible for the Authority and its member governments to resist what the District felt were necessary cutbacks.

Harrell had Farrell put the first draft of this proposal together in October 1996. There was a problem, though. The local governments most certainly would not, could not, give up their costly infrastructure, including pump stations, treatment plants, pipelines, and all the other elements that get water to people. They'd paid for those costly facilities long ago, which turned them into revenue sources today. Replacing them at present cost would be an expensive proposition. What incentive would they need to hand them over to the Authority? Was that even possible? Harrell's ideas were always big. His goal was to create a strong regional utility: something the District had insisted would be a more workable management structure. For the District's part, it would be much easier to have one large permittee for public supply, rather than having the separate local governments.

***

I saw the first draft of the plan in November 1996. I hated it. I was furious with Roy and accused him of putting money in his friends' (the local governments') pockets. The plan included taking money from Save Our Rivers and Preservation 2000, programs designed to protect critical habitats in the region. That money shouldn't be used to pay off local governments for overpumping. It was their responsibility. "If you just hand it to them, then what?" I asked Roy. "I won't help you with this," I told him. "This is wrong. All the people who helped us, all the environmental groups will hate us. And they should."

I railed at Mark and Roy because I saw this as the beginning of the end for the District. We would lose the public support we had fought to deserve. We didn't *deserve* their support with this stupid plan. When I had Mark alone, I confronted him.

"Why?" I asked. "You know what's going to happen here. Why don't you tell him how awful this is?"

"Look, Roy asked me to draft this plan before I left," Mark told me. "I don't have the strength to fight him. He's committed to this. Honey, I'm outta here in just two months. It's not my fight anymore."

In truth, Mark didn't like it any better than I did. "It amounted to paying the local governments for being bad actors all these years," he said.

Mark and Pete were already gone. I was alone. With Roy.

***

The plan went through a few more drafts, but by December 1996, the Governing Board and Roy Harrell were tied up trying to find a replacement for Peter Hubbell. The new executive director would find a replacement for Farrell. While they were still physically at the office, they were essentially gone. Three months was too long to hang around once you've decided to go, I thought then. It was too long for them, and it was too long for the District.

Ed de la Parte maintained that no settlement between the District and Authority—let alone the member governments—would be possible with Hubbell and Farrell in place. But many others disagreed, including Ed Turanchik and Mayor Fischer. Like de la Parte, the two had their detractors.

"I thought they were great allies," Turanchik said.

"It didn't matter by that time," Fischer said. "We were trying to figure out what to do with ourselves internally [at the Authority], so we weren't really focused on the District and its issues."

"The District's power was waning," said Manson. "They peaked and maintained their power for a long time, but people were working to drive a wedge between the District and the legislature. For a long time, that didn't work. Even at the end, when the member governments went to the legislators, they were already talking about a central utility with a uniform rate. The District had already been there."

The District was working hard on an alternative solution, but with tremendous internal uncertainty, it was impossible to advance the agenda. The Partnership Plan went into a semidormant period as the task of selecting a new executive took priority. Political infighting between Deputy Executive Directors David Moore and Rich McLean was hurting the agency. Hubbell and Farrell were lame ducks, unable to control either of the deputies.

"I tried to tell Dave to be cool, don't play games," Farrell said. "That lasted for about a month."

By late October, even staff not part of the inner circle could see the campaign underway to replace Hubbell. Each of the deputies had their board supporters; each had their detractors. All of them influenced the agency. Activists campaigning on behalf of their favorite candidate made the problem worse. Most of them wanted McLean to get the job. They lobbied board members to show their support. They were so involved with District staff that they thought they could influence the process. Ray Campo, a board member known for his "flexible" positions on issues, meaning he seemed to have no baseline values, was McLean's staunchest ally. Richard Owen, longtime District planning director, originally applied, then withdrew. The District was adrift with no center and no clear focus. The speed at which it happened was stunning.

But McLean and Moore were not the only candidates interested in the executive's job. Two others already in Florida and known in water-management circles were hard at work campaigning for the job. One was an obvious choice, the other a complete surprise.

John Wehle, the longtime assistant executive director of the St. Johns River Water Management District, immediately expressed interest. That was no surprise. Insiders thought that it might be Wehle's turn. He'd done well for St. Johns, consistently delivering on its political agenda in the Florida legislature. Some District supporters thought he'd done too well, stalling and modifying our Local Sources First effort, among others. The other candidate was a complete surprise—Sonny Vergara, general manager of the Peace River Manasota Water Supply Authority, the guy who walked out on the St. Johns River Water Management District so many years ago.

"There was an interest, a desire, a compelling emotional drive to do this," Vergara said. "I asked a lot of people. I asked them if they thought I could do it, if I could make a difference. And I got a lot of encouragement."

During this time, the District had hired communication consultants from the University of South Florida who were interviewing various staff. The issue of competition between the in-house candidates was evident even to them. It was coming up in interviews.

When the consultants provided a draft report, they noted the issue of the competing deputies, but in vague terms. On December 20, 1996, in a special board meeting to evaluate candidates, Harrell waved a copy of the report at board members, telling them what was in it and why they needed to choose new leadership—from outside the agency. The communications tool that should have helped the District get a grip on itself was instead being used as a political weapon against internal candidates.

The executive recruiter moved through the list of candidates, briefly summarizing his thoughts about them based on his preliminary interview. "Sonny Vergara, he's from Brooksville, he manages a small organization, but he's personable. For major management experience, you'd have to look way back. He's very well connected. He knows quite a few people. I've gotten a few letters in support of him," the recruiter said. Flipping pages in his notebook, he continued, "He said he resigned [from the St. Johns District]. He hired a number-two person, the board didn't like the person, and there were some issues. . ."

Sonny's view of the District was that it needed to be more proactive, not just a regulator, the recruiter concluded.

Ray Campo asked, "What was the reason he left?"

"He decided to move on, but there was political pressure," the recruiter answered.

On the other hand, the recruiter said, "John Wehle comes from a similar-sized organization. My sense is he spends a lot of time in Tallahassee." Several board members chuckled. One said, "Your sense is correct."

Someone asked, "Did you get letters on him?"

"Yes, quite a few. He understands water-management districts," said the recruiter.

"You talking about Vergara?" a board member asked.

"No."

"Did you go over Vergara?"

"Yes."

"I must have missed it," said one board member.

"You didn't miss much—it was a short résumé," said another.

"Behave yourself," said a third, lightly.

"That's what I heard," said the second board member.

The discussion moved on.

In the end, the board asked the recruiter to provide his recommended list of candidates for interview. In alphabetical order, he offered six candidates:

- Alan Ispass (out of Florida)
- Rich McLean (deputy)
- Dave Moore (deputy)
- James Morris (out of Florida)
- General Paul Vanderpoog (retired military, head of Hillsborough County Utilities)
- John Wehle (St. Johns District)

"If the board is willing to do eight interviews, I would add Vergara and Vang," he said.

Board member Jim Cox (the same one who had called attention to Steve Monsees's presentation three years before) said that the board should "promote from within unless an outside candidate is clearly superior." He said that morale and the issues dictated the necessity of hiring from within.

Former chairman Joe Davis immediately talked about an outsider, acknowledging that "there may be someone outside that may be outstanding. It sometimes shakes an organization out to come from outside. It can be a hassle, but you get new blood, new thinking."

As chair, Roy Harrell addressed the group. "One of the concerns I've heard is a conflict we've had inside the District. It's going to come out by a professor that we've had looking at the organization. It's going to come out that we've had a lack of good communications in our deputy executive directors. It's occurred. It's unfortunate. I think there's been a level of competition."

He turned to Jim Cox. "To some degree, Jim, there is destructive competition among the deputies. And I must tell you that I would tend to go in your direction, to go inside. I must tell you, I'm really going the other way now because I think what you need is that the District work in harmony. I'm concerned. Sometimes bringing in someone from outside—there's a healing process in that. It's just a point. I'm not citing facts; it's just occurred." Roy suggested that Vergara be interviewed. "Vergara," he said, "he's gotten done what needed to get done. I think that part of his cool style has helped there."

"I felt sick," Gilliam said. "He told the Governing Board that the District was in a bloodbath from the competition between Rich McLean and Dave Moore. What he did was disgusting," she said, angry still.

That day, board members went through the list of candidates, accepting some, rejecting others. They decided that interviews would be held at the Tampa Service Office during the Martin Luther King holiday in January 1997.

While the District was rocked by events, the Authority wasn't doing well either. Issues were festering behind the scenes, and member governments were generally unhappy. Jerry Maxwell, the general manager, had not been a particularly public figure since his arrival in 1995.

He focused on developing a plan for new water supplies. To do so was a politically tricky matter. Pasco County wanted no part of providing new water supplies to the region, figuring it had done its part. The District was calling for an exploration of desalination, which Pinellas was dead set against and which he had declared would happen over his dead body. Tampa was pressing for an indirect-potable project, the "poopoo project" as opponents dubbed it, and Pinellas continued to want new or expanded groundwater supplies. Maxwell was walking through a minefield and not evoking a lot of sympathy.

"People targeted Jerry as the problem," de la Parte said. "It's part of the legacy of demonizing people in the late 1980s, early 1990s," he added.

It may have been more than that. Jerry Maxwell has a talent for saying things in such a way as to allow the listeners a wide range of interpretation. It's called strategic ambiguity, and he was very, very good at it. Some Authority board members exposed to this for eighteen months began to see Jerry as duplicitous. At times they felt intentionally deceived. Maxwell denied it and pointed to the staff closest to him. "The people who know me best are the ones who trust me the most," he claimed. "Because they know I'm out front, leading them." He did indeed evoke tremendous loyalty from the senior staff.

But ask anyone beyond that inner circle, and you'll get a different take on who he is. There was an arrogance about him, his critics said, and he was no consensus builder. Maxwell was anything but out front, according to his critics. Still, he was managing to do something others had been unable to do for some time—identify water-supply projects for development. Many believed that in doing so, he was advancing the Pinellas agenda. If that were true, there would have been no reason for Pinellas commissioner Rainey to leave. The activists hated Jerry Maxwell, and the member governments were suspicious, but there was no alternative. When Jerry was hired, he was given a contract with a guaranteed buyout if he was terminated early. Lifetime public servant and District deputy executive Eugene Schiller said that most public administrators leave the same way they come in: "fired with enthusiasm." It seemed that Maxwell was perpetually heading in that direction, and yet, given the state of negotiations over reorganization and the District's money, it was hard to think about bringing in someone new. There was no one on staff to replace him. The thinking was "Since we have to pay for him, we might as well keep him." And yet every year, his contract was renewed and extended. Although the member governments complained about him, they continued to affirm the job he was doing. They continued to give him raises.

"He played strongly to whomever he was talking to at that moment," Ed Turanchik admitted. "He could weave a version of 'facts' that sits well with that person. But, you know, he would have five different board members with five different versions featuring colors and gradations that are just slightly different," he concluded with a big, approving smile on his face. "He was a bit of a conniver. But we needed that right then. The deal could not have been brokered without Jerry Maxwell. He kept the process moving, and we hit our benchmarks."

"Well, it took a lot of skill sets to pull it off, you know," said Seibert with a shrug.

Turanchik admitted that he knew what Maxwell was doing but added that having a plan and the Authority reorganized was worth a bit of treachery. "It's kind of like the parents of a teenage daughter," he said with a big laugh. "She's gone off to college, and she has a new boyfriend, and as a parent, you know what she's doing, but you don't want to hear about it. You don't want her to come home and tell you how great it is. That pretty much sums up my feelings about Jerry Maxwell at that time. I didn't want to know how he was working it; I just wanted him to get it done."

Whatever issues surrounded Maxwell at that point, it was clear the Authority board wasn't about to make a change in management with a possible reorganization looming. He was going to be there for a while. As it turned out, his unique qualities might just have been those necessary to make water work, to get the job done, including reorganization.

Still, it was St. Petersburg mayor Fischer who pointed the way to reorganization. "We were having a lot of good workshops," he recalled. "We had our city managers or administrators with our utility directors and us. We were sitting around a big horseshoe table, and you could just see some softening going on." Mayor Fischer told them, "Look, we have to put together a real water authority. The city's spent $2 million on litigation already, but it's not getting us any new water. There must be a better way to do this."

With that, the mayor went to the whiteboard at the front of the room.

"Look, let's pretend that we're Florida Power," he told them. "When we flip on our light switch, we don't care if the power comes from somewhere else or from the power plant next door. It's their hassle to site the plant, not ours. It's their job to see to it that we have electricity. They don't change the price each time they build a new facility," he said. "And they don't have one price for electricity from one place and another from somewhere else. They develop new plants as they need them, and they charge everyone the same rate. We can do the same with water. We can all turn our wellfields over to the Authority and let them be responsible for getting the water supplies. We'd have to be in it, not as a coop, but as a real utility. The only way to get out is if the utility fails to deliver the water."

That was the genesis of the solution, and they could all see it, but could they achieve it? The District had been talking about a strong central utility for some time, but Mayor Fischer, as an insider, had articulated it as a real possibility. No longer a coop but a real regional utility. The Authority would not have a single member who would be able to stop water-supply-development projects. It could correct a structural problem resulting from the formation of the Authority—a "problem" that had effectively allowed Pinellas County to control water-supply development. As a central utility, with a single water rate for each local government, it could move ahead with projects based on a majority vote. Some members would pay more, some would pay about the same, maybe less, but they would all be dependent on a single source—an outside source—for their water supply.

"The day we had that meeting, I knew that we could fix this," Hillsborough County commissioner Turanchik said. "I believed we were really going to do it."

"Turanchik and I struggled with the details," said Steve Seibert. "Dave [Fischer] had the big picture from the beginning." And yet it would be more than a year before the group accepted either the premise or the terms.

In short, the faces changed, and when they did, the possibilities changed. The shared sense that "we're all in this together" replaced the "I got mine; good luck getting yours" attitude. Principled people, together, focused on doing the right thing. The District's funding made a difference, too. But it started with people. People that wanted change.

Chapter 16
Unchartered Waters

The campaign for the executive director's position moved forward rapidly. Roy Harrell's comments regarding "going outside" the agency had little effect on the behavior of either internal candidate. An executive recruiter reviewed résumés and short-listed candidates for interviews. The deputies jockeyed for position on the shortlist and the even shorter interview list.

"Wehle and I were involved in an actual campaign," Vergara admitted. "Once I started to contact people and ask for their support, it wasn't too long before I ran into those Wehle had already visited." Vergara said he'd wanted to get to people first so that they had him in mind. "I wanted to get support first and remove the opportunity from others," he said. When Wehle had already been there, "people felt conflicted; they liked both of us, but they weren't as clear as those I visited first." He didn't stop with phone calls and visits, either. He asked for letters, and they came in.

"Once I ran into John Wehle while all this was going on, and he said something like 'Here comes the traveler' or words to that effect. And I said, 'What?' He said, 'Yeah, man, you're the traveler because every time I go in a door, you've already been there.'" Sonny smiled at the memory.

"There was a campaign going on," he continued, "but I don't think it had any influence on the board's selection. I thought Wehle had collected a lot of baggage in Tallahassee."

Wehle had been in Tallahassee doing legislative and political work for the St. Johns River District while Sonny had been at the Peace River Manasota Water Supply Authority and out of harm's way. Vergara was getting rid of baggage he collected over prior years.

***

Only the Governing Board members, selected staff, a handful of water activists, and the candidates were present for interviews. I was there to send the news release afterward and make the media calls if they made a decision. The interviews were held on Monday—Martin Luther King Day—and Tuesday. Monday was a holiday, so no one was around. No media. No staff. It was a public meeting, as were all Governing Board meetings; even so, only a handful of staff were present on Tuesday. It wasn't that they'd been told they couldn't come, but tacitly people were expected to be elsewhere. Only Linda Sullivan, I, and a couple of others attended. This was a decision that might change the entire direction of the agency, it was a public meeting, and virtually no senior management were present? It didn't take a scientist to figure it out. I was in and out of the room most of the day. There were only a few presentations I really wanted to hear, including a guy from Orange County Utilities and another from the Tennessee Valley Authority. I also wanted to hear John Wehle. Those I didn't care to hear: the staff candidates Rich McLean, Dave Moore, some guy from the Carolinas, the retired general that headed up Hillsborough County Utilities, and Sonny Vergara. I thought I knew what Rich and Dave would say, and I knew Sonny, too. None of them seemed likely to get the job.

Before the meeting began, I spoke briefly with former chairman Joe Davis inside the executive office. "Be careful," I said to him. I'd heard that individual board members had been told by a legislator to "pick a Florida resident." I thought for sure that meant Wehle. I didn't know him personally, but I'd heard he was "very political." I had vague memories of Wehle working against Local Sources First in Tallahassee. Nothing personal, of course. Just a policy issue. Still, we got this far by "being with the people" and not by "being political." Hubbell had good political instincts, and for the most part, we operated in a world where he was the only one that had to consider the politics. We thought we practiced the politics of what is right. Wehle was a player. Wehle had experience. It was Wehle's turn. "Not Wehle," I was thinking.

Of the candidates interviewed, one was clearly Pete 2.0. I met him before the interview. He looked like Pete, he sounded like Pete, and I think he even had Pete's low-key demeanor. He came from the Tennessee Valley

Authority. Linda Sullivan and I looked at each other. There was hope. There was another Pete. Even his answers reflected Pete's style. Thinking about it now, I think his answers reflected Pete's political instincts. Still, Linda and I found it eerie and exhilarating. This was an excellent development. Surely, the board would see what we saw.

When John Wehle came in, there was something about him I immediately liked. Solidly built and not very tall, he carried himself with assurance but not arrogance. Before he came in, I expected something else. He had silver hair and a nice smile and was well dressed. Attractive. He looked like an executive director. When he began to speak, I realized that he sounded like one, too. I measured every answer against "Can I work for this guy?" I loved my job at the District, and I knew that I was probably leaving, but I really wanted to stay. I wanted the board to choose an executive I could work for. I walked in thinking that was hardly possible, but now, I dared hope that I might be able to stay. To my surprise, I liked John Wehle very much. He seemed honest. He was articulate, and he was quick. He was funny. This had definite possibilities. It was Joe Davis who asked the question on everyone's mind.

"John," his gentle southern accent masking a rebuke in the form of a question, "you have a reputation for being very political, sometimes controversial. How do we know if we hire you that our policies will be carried out?" I assumed that Joe's comment related to our alternative-sources policy, one opposed by the St. Johns District.

"I understand," he said to Joe. "I've worked very hard and very well for the St. Johns District. Sometimes that makes me a lightning rod. I've been effective, and I've been the front guy, so I've taken the heat. If I worked for you, I would fight just as hard for your policies."

Yes. Things were looking up.

I was largely out of the room for the Dave Moore and Rich McLean interviews, though Linda took notes for me. It seemed both had some definite ideas of how to "fix" the District. I'm glad I was out of the room for most of it. I liked both very much and didn't want to hear them say we were broken. I thought we were just fine. I also missed the presentation from Sonny Vergara, but Linda told me it was nothing special and seemed like a courtesy interview to her.

It was getting late. When the balloting was all done, Sonny Vergara was the new executive director of the Southwest Florida Water Management District. I was stunned. Hadn't the board members heard the interviews? I stepped into the hallway to find an empty office space. I needed to call reporters and tell them. Mr. Martin, a board member from St. Petersburg, Roy's best board friend, was there. I gave him a hug.

"I can't stay, you know," I said to him.

"Why?" he asked. He seemed surprised. I was too tired to get into it, but I knew, quite clearly.

"It just won't work," I said. I went into an office and began to make the phone calls to set up interviews for Sonny. On the way home, I called Mark Farrell.

It was three years later when I was told by a board member, in confidence, that a call had come directly from the governor's office. "Not Wehle," the caller had said. That limited the alternatives. "Boards do things, they just do," Mark said, trying to comfort me.

***

"Rich was the only person interviewed with any vision," Gilliam insisted. "Dave kept saying, 'I'm a team player, hire me, hire me,'" she said, ever-present sarcasm creeping into her comments. "Being a team player is not being a leader. You needed a leader for the District, and you didn't get one."

Some said Sonny's good friend since boyhood, Jake Varn, pulled in every political favor he had to get the job for him. Some said Roy Harrell sold out (they didn't say to whom or for what) or caved into political pressure, that the fix was in, and so on. None of that mattered. Mark Farrell said it:

"Boards do things, they just do." He was right. This board had done a thing; it had hired Sonny Vergara. Board members knew him. He'd been around a long time. He had a long and public history. One on one and in small groups, Vergara exudes warmth. He's got a good sense of humor, and, most days, he could poke fun at himself. Sonny is a likable guy. But Sonny was going to change everything.

He made no pretense about who he is, how he worked, or why. His selection as the new District executive director was generally met with support by outsiders. District staff were a different matter, though many took a wait-and-see attitude. They didn't have another choice. Peter Hubbell had been the executive director for nine years. It was hard to imagine that Sonny Vergara—or anyone else—could take his place. There were two things the staff knew about Sonny. He had a bad temper—enough to make him walk out on the job of executive director at the St. Johns District—and a long memory. We didn't know he didn't much care for us. But he showed us very early on.

Wishing for Water Wings

A printed email clipped to the copy stand at my home desk read:

Subject: Thought for the day

Date: Wednesday, January 22, 1997, 19:23

From: Mark D. Farrell

To: Honey Rand

"When all else looks dim, remember we still love you. The Boys."

I don't remember why he sent the note, but I was glad he did. It was the day after Sonny's selection, the day it appeared in most of the papers. Mark was the only person who really understood how sad I was and that I was only staying for Roy Harrell. As chairman, he had his hands full with the Partnership Plan, a new executive director, the upcoming legislative session, and all the other issues facing the District. We didn't have much time.

Sonny started work right away while the Peace River Manasota Water Supply Authority searched for his replacement. I called Sonny in the evening to brief him on communication developments and issues. He was the executive director, and he needed to be able to guide the agency's public tone. Might as well start sooner rather than later.

I remember one of those early phone calls. I had a list of issues to cover with Sonny, and for some reason, he stopped me on one item. I don't remember what it was, but he said this to me: "You people are like the royalty in England. You've been inbreeding for so long you can't even tell the difference when something is wrong. What I'm going to do is throw open the doors, let down the drawbridge, and bring in some fresh air and new blood."

I was appalled on two levels: first, that our new leader thought such a thing, and second, that he would say that to me, the official keeper of the public image. It was only a few weeks later that he walked into the Monday-morning senior-staff meeting, his first, and said to us, "People, people don't like you, and if you don't know it, you need to get to know it." We looked at him, we looked at each other, and we said nothing. It wasn't a matter of whether I would stay, but how long I could stand it.

"Had to. Had to. Had to do it," Sonny later said. He smiled, remembering that first meeting with the senior staff, most of whom he'd known professionally, some for decades. He'd seen our public-opinion numbers with off-the-chart public support, but to Sonny, it didn't matter. His constituency was not going to be the public or environmentalists. He made a different choice.

"Public-opinion polls go to common households and people who care about the environment and water supplies," Sonny said. "They probably felt some sympathy for the purpose of the water-management district, but it didn't matter, it doesn't matter. There's a whole different society out there called local government, legislators, opinion leaders in the community, and they're the ones ready to cut the District off at the knees about neck high."

As senior staff prepared for changes to come, the Authority decided the Governing Board was pushing for too much control in specifying water-supply projects. It didn't like it. It decided to make its feelings known.

The Price of Peace

Hillsborough County commissioner Ed Turanchik appeared before the Governing Board in January 1997 in his capacity as Authority chair and told it its money was welcome but its direction was not.

Roy Harrell remembered the main message being "We'll make the decisions about what sources to bring. You provide the money."

"The District needed to keep a dividing line between the role of regulator and the role of supplier," Turanchik explained. "We had a good thing going at West Coast [the Authority]. We were moving forward."

The idea that the District would once again be a partner and not an adversary was a big relief. Mayor Fischer felt like things were coming together for a positive change. It took some time and tremendous effort, but change became the prevailing idea. The only problem was: what change and what timing?

Because of Turanchik's visit, the District made two critical modifications in the proposed partnership. The first was to recognize the Authority's independent ability to decide what water supplies to bring online. The District could and did insist that its money be used only for alternative supplies. Under Hubbell and Farrell, alternative supplies were defined as those that were safe, sustainable, environmentally friendly, drought-proof, and cost-effective. Under the new regime, river water was an alternative supply. It was an alternative to groundwater, yes. But it was still rainfall dependent. To activists and many District staff members, this was a fundamental shift in policy. Previously, the District had insisted that its money be used on technologies like desalination, aquifer storage and recovery, and wellfield or wetland rehydration.

The second modification was to allow the money to be used for pipelines. Under Roy Harrell's direction, the board and the staff were beginning to understand that cutbacks, if they were to come, would have to coincide with new water supplies. New water supplies would be expensive. Fortunately, the District had the money to invest. Joe Davis's program, the New Water Sources Initiative, had a lot of money and would have a lot more in the coming years. The board just needed to agree that instead of a challenge-grant program, the funds could be directed to the Authority.

The new executive at the District was a water-supply guy. Hubbell and Farrell had been regulators. Hubbell was a hydrologist, Farrell a professional engineer. Their perspective on policy was much different from someone who'd been developing water resources for ten years. Folks at the local water authority, especially the utility directors, thought that having one of their own as District executive director was a welcome change. Given the changes on pipes and policy that came right away, the District's supporters weren't sure at all.

Prophylactic Politics: Protect Yourself

With two water-supply guys at agency helms, one would think that détente would come quickly. It did not. The legislative session, scheduled to begin in March, promised to be as combative as ever. It was as if the Authority, sensing the District was in flux, was pressing its advantage while Vergara was assessing the political landscape. Both Vergara and Maxwell understood the pressures and the pitfalls of water-supply development, but they were personally very different.

"Jerry is very smooth," water lawyer Ed de la Parte observed, "while Sonny wears his emotions on his sleeve. Jerry is cool, he's controlled, where Sonny, if he's ready, will just go after you. Once Sonny's anger has blown through, that's it. With Jerry, well, it's hard to tell."

It was clear the Authority was going to make a move in Tallahassee to take, without strings, a portion of the District's ad valorem funds. To make its reorganization work, it needed money. The District's Partnership Plan offered to fund new water supplies, but it didn't sit well with the Authority for several reasons. For one thing, the District was using the funds to advance its alternative-water-supply policy. The Authority resisted. Its move on the District's ad valorem taxes during the session was intended to take the money with no strings attached. A meeting at the Tampa Service Office was scheduled to discuss the strategy for Tallahassee and the communications necessary to protect the District plan and advance Roy Harrell's Partnership Plan. In attendance were most of the executive staff, including Deputy Executive Directors Dave Moore, Rich McLean, and Gene Schiller, General Counsel Ed Helvenston (acting as the assistant executive director), Sonny, and Roy.

Harrell began with a brief overview of the latest modifications to the plan. He talked about the rumors of fund reallocation from us to the Authority, and he opened the discussion on the strategy for Tallahassee and the likelihood that the Authority would succeed. He was obviously concerned, as was Vergara.

***

At some point, Roy turned to me. "What can you do?" he asked.

"Everyone is so ready for this problem to get solved, I think they have a good chance of getting our money unless we provide a logical alternative to it," I told him.

"What do you mean?"

"I mean the legislature may not want to go through the hassle of allocating our money to them, but they need an excuse not to. We need to push the Partnership Plan as an alternative," I suggested.

"No," said Sonny emphatically. "It doesn't need publicity."

Roy turned to me. "Is there something else we can do?"

"We have two choices," I said. "We can either positively advance our plan, or we can take theirs apart."

"No," Sonny said. "I don't want the Partnership Plan out there now."

"Fine," I told him. "I don't care which direction we take. I'm just telling you the alternatives. We either need to give legislators something, or we need to explain why the Authority's master water plan won't work. Otherwise, the only thing between where we are and solving the problem is the District's money. And everyone wants to solve the problem."

"No," Sonny yelled at me. He stood, and he pointed at me, shaking his finger. "I'm sick of your negativity," he said. "People hate you."

I was stunned at the outburst. "People hate me?" Like, what was his point? I didn't know whether to be angry, embarrassed, or amused.

"You've got to stop all this negativity," he said. "People hate you," he repeated.

Roy's face reflected his unhappiness at the exchange. Looking down the table at Sonny, he asked, "People hate Honey, or people hate the District?"

"The District," Sonny said. As the momentary anger passed, he sat down.

"Who hates us?" Roy asked. "Utility directors? Who cares? People don't hate us. You need to see our public-opinion numbers."

But Roy didn't know that Sonny didn't care. On one level, I was glad for Roy to see what senior staff were seeing. The Governing Board had hired an executive who seemed to have no regard for his senior advisers. One who still could not control his temper, just like so many years ago at the St. Johns River District board meeting.

At another level, I wanted out, but I couldn't leave Roy.

***

Sonny didn't remember the exchange, but others at the table did. Vergara said the sentiment that people recalled him expressing was valid. He clearly remembered the intense negative feelings directed at the District.

"It's not that I had no regard for the senior staff," he said, "but they had to know that change was at hand. They were blind to their problem."

On the Partnership Plan, we decided to defer the decision about what to do until there was a more accurate reading of the legislative mood. Before the session ended, a briefing was prepared, and the District was out giving presentations on the Partnership Plan. I made more than a hundred presentations to social and service groups, chambers of commerce, and neighborhood associations between May and September as the District offered its plan to develop alternative water supplies through selective investment in them.

Money Counts: The Battle for Control

"There was a proposal to direct some of the District's ad valorem to us," said Authority general manager Jerry Maxwell, straining to remember, "but I didn't think it was a good idea." He took a call from Authority lobbyist Peter Dunbar. "He read me a list of possible ways to get the money we needed for the water-supply projects," Maxwell said. "That was one of the options, but we didn't pay much attention to it. It would have been too much trouble."

However, others, including the Authority chair, Ed Turanchik, clearly remembered a move to "take" money from the District. The question that no one answered is, why? Why, when the District was working on a Partnership Plan, try to "take" the money? Power and control? Get the money with no strings attached? Just to prove you can?

"It didn't come from Pinellas," de la Parte said flatly about the move to take the money. "It may have come from Senator Latvala or Jerry. Maybe it was Turanchik?" Ed speculated the thought may have had something to do with the District's taxing powers. Had the District reneged on the Partnership Plan, it couldn't be sued to raise taxes, so down the road, there would be no way to force it to provide funding. Still, that was a stretch. Given the political context, it's hard to imagine a scenario where the District would refuse to provide money to underwrite the water supplies agreed to in a historic partnership.

"Latvala tried to frame a solution legislatively," Turanchik recalled. There were several attempts, and there was the ongoing threat that if the region couldn't fix its own problem, the legislature would fix it, just like in the 1970s. When Latvala first became involved in water issues, Turanchik and others were afraid he would simply push the Pinellas agenda. Instead Latvala rapidly became a force for positive change. "Eventually, I agreed with Jack's vision," Turanchik said. "But it couldn't be done legislatively. It had to be our solution. It couldn't be imposed. He and I had some very heated exchanges over what needed to happen."

"Latvala is his own man," Vergara said with admiration. "He was a shrewd politician, and he had a lot of guts. A lot of courage." Vergara agreed that Latvala had expended a lot of political capital trying to bring Pinellas County around. "He was willing to do that kind of thing. That's why Latvala is Latvala."

Latvala being Latvala, he was never afraid to push an issue. So when Hillsborough County was pushing very tough minimum-flows-and-levels standards, Latvala placed a call to Ed Turanchik. Hillsborough wanted tougher standards. Pinellas didn't.

"Latvala called me up and accused me of trying to sabotage the reorganization process," Turanchik remembered. "He threatened to get into Hillsborough County's management and growth plan. I told him he didn't have the power or the prerogative to do that," Turanchik said. "He forgot that he was talking to someone who knew the difference."

Turanchik and Latvala were ultimately working toward the same goal, although it took a while for them to understand it. Momentum was moving in the right direction. With Latvala's help, Pinellas County was running out of alternatives.

***

Toward the end of the legislative session, Vergara returned from Tallahassee with a message for me: "You're immortal." Though not officially naming it, lobbyists referred to an amendment tied to the water bill as the Honey Rand Amendment. From here on out, water-management districts were required to produce annual reports detailing their expenditures for public communication. Some say the District's communication program was too good. With so much public support, the District made some legislators uncomfortable. Deputy Executive Director Eugene Schiller proudly said he annually puts zero dollars in the category for public relations. "We didn't do public relations," he said. "We did public education."

***

Just after the session ended on May 15, 1997, Pinellas commissioner Steve Seibert sent a memo to his colleagues on the Pinellas Commission. "Five Questions and Three Options" was a summary of his heartfelt concerns and best analysis of the dilemma in which they found themselves. He called water their "most technically and politically complex issue," and he noted that the memo was prepared with "respect for the natural environment, which every member of this board cherishes."

His questions: Will our current wellfield facilities be cut back? Will we fight those cutbacks? Will we develop other water-supply resources? What will they be? How will we pay for them?

His options: Continue as we have in the past. Concede some of these legal rights and proceed with implementing the agreement in principle. Go it alone.

He borrowed from Robert Frost in "*The Road Not Taken*" "*Two roads diverged in a yellow wood, and sorry I could not travel both and be one traveler, long I stood and looked down one as far as I could to where it bent in the undergrowth.*" "We are looking down that road as far as we can to see where it will take us," he wrote to the other commissioners. "It is difficult because the paths do bend in the undergrowth. I have looked down the path we have been taking and believe we need to change direction."

He continued:

> The most difficult impediment to resolving these disputes is that we do not communicate well within the region. The Pasco and Hillsborough commissioners have seen lakes and wetlands dry up near our wellfields and are upset about it. Pinellas' position is at least as passionate. We have always pumped within our consumptive use permits and feel the drastic regulatory restraints upon us lack scientific justification. The numbers are arbitrary, fueled by politics and inequitable in the sense that neither agriculture nor industry is being restricted the way public supply is being cut back. We have a long commitment to environmental protection, to extensive and expensive conservation measures, and are being condemned not by the facts, but by a slicker public relations effort. We cannot proceed with solutions without more effective and honest communication.

"It was time to make a choice," Seibert said, reflecting on the letter and proud to have put it on the table. "And they did. They gave me the Authority to go ahead and cut a deal." He saw it as a major turning point for the Pinellas Commission. It was also a major turning point for the region.

"A time was coming," he speculated, "when someone was going to take a real look at all the costs attendant to this, not just Eddie de la Parte. It wasn't going to be pretty. Someone was going to figure out that it was public dollars, fighting public dollars, fighting public dollars, and when it was all added up, the District would be there, too. Ultimately, it would have gotten everyone."

The Ladies of the Lakes Become the Water Bugs

By 1997 many of the water activists were understandably tired. They were sick of the ongoing negotiations that never seemed to get anywhere, and they were acutely aware of the shifting power structures around them. Worse, with Hubbell and Farrell gone, they felt they'd lost their strongest allies. The frustration was evident in some of their remarks.

"I made board members angry," Gilliam Clarke said, "but I don't believe in getting up and haranguing them, I really don't. It's counterproductive. I don't agree with them, and I might step on their toes or butt them a little bit, but I'm not going to hurt anybody."

Gilliam's attempt to "keep things fresh" was sometimes stinging. She hated the idea that Roy Harrell and Sonny Vergara were working so closely with the Authority. She was afraid the ground gained with (political) blood, sweat, and tears would be ceded to make a deal. She was afraid that Pasco County would be sold out.

"At public comment, I told the Governing Board, 'You might notice that you never see Jerry Maxwell here anymore. That's because the chair and the executive director are always down there begging to please cooperate. It reminds me of *The King and I*: "Yes, Your Majesty. No, Your Majesty. Show us how low to go, Your Majesty. Don't let us up off our knees, Your Majesty. Give us a kick if it please, Your Majesty. Give us a kick if you would, Your Majesty. Oooh! That was good, Your Majesty,"'" she concluded.

"The room was in stitches," Gilliam said with a giggle. "They were rolling on the floor. That hit him. He was furious." She was talking about Roy Harrell. Gilliam saw the District's position as weak. She blamed Harrell as the chair and as the representative of St. Petersburg and Pinellas. It was just as Joe Davis had predicted. Despite his best intentions, Harrell was accused of advancing the Authority's (and by extrapolation, Pinellas's) agenda. The only laughter in the room was in Gilliam's mind.

"The District became a shadow of its former self," Gilliam said. "The District went from being a star in water management to being, well, it's hard to talk about it. The District is an inferior adjunct of the Authority."

"There came a time when activists were repeating themselves," Manson observed. "When they did that, even if what they're saying was important, accurate, and true, it got old because it was old."

"Activists must deal in facts," Gilliam believed, "even though facts can be wrong."

Gilliam was occasionally known to deal in what might be called today "alternative facts." But there was one fact upon which most activists agreed—they had no use for Jerry Maxwell. Even the balanced Eileen Hart had harsh words for Jerry. They variously thought he was working on the Pinellas agenda or his own. For whatever reason, elected officials may have had issues with him, and the water activists held him up as the Antichrist of water-supply development. His detractors were vehement. He was variously incompetent and Machiavellian. Gilliam was convinced that Maxwell represented Pinellas interests.

"I've sparred with him in public meetings," she said. "I'm opposed to him. He's incompetent. He's a political animal and a kingdom builder, and he'll stop at nothing to build this one for himself."

Yet it was Maxwell who consistently tried to make water work, to bring new projects online. The longtime activists didn't like the master water plan. It was too dependent on rainfall.

"The master water plan was little more than smoke and mirrors," said Eileen, reflecting the feelings of many water warriors who felt the plan was too heavily rainfall dependent.

She wasn't necessarily against new water-source projects or projects dependent on rainfall. She understood that until replacement supplies were online, Pasco County and northwestern Hillsborough County would continue to be pumped beyond reason, and she was certain that Maxwell couldn't care less.

"The biggest thing was, they—the activists—called for cutbacks to wellfield pumping, and we were trying to make it happen, and then they opposed the new water projects," Turanchik said, exasperated.

"Jerry is in everyone's backyard," St. Petersburg mayor Fischer observed. "People don't like projects being built next to their homes, they want to know what it's going to do to them, and Jerry is the guy building the projects. If it's desal, he's in Tampa Bay; if it's a new wellfield, he's putting a straw in the ground; if it's the Alafia or the Hillsborough Rivers, he's executing the plan—a plan that took thousands of hours, that was approved by the board. Then the board changes, and that's a disadvantage."

Many water activists fought the agreement that reorganized the Authority to implement the master water plan. Suspicious because they were largely left out of the process, and with little faith in the District's leadership, they were opposed. They supported the Partnership Plan, the idea that the District would have some control over new projects, but believing the District was ceding control, their worst fears were realized.

"The District used to listen. They used to try to do something," said Clarke. "That disappeared. The District does nothing now." Her tone was uncharacteristically flat.

The proposed "new" Authority was chartered to have nine members, two from each county and one from each member city—that is, Hillsborough, Pasco, and Pinellas Counties and the cities of Tampa, St. Petersburg, and New Port Richey. In the reorganization, Pasco County's largest city, New Port Richey, got a vote. This changed the balance of power somewhat because New Port Richey, located on the Gulf Coast of Pasco, shared more interests with Pinellas County and typically sided with it, even when that decision was harmful to central or eastern Pasco residents. The activists had counted votes enough times and knew that two votes from Pinellas, one from New Port Richey, and one from St. Petersburg would generally be together—all it took was one Tampa vote for Pinellas to regain control of the Authority.

In the past, Tampa and, occasionally, a Pasco representative voted with Pinellas, so it was not a distant possibility. The activists were afraid of losing ground. The days of solidarity among the District, Hillsborough, Pasco, and the activists had passed. The COLA didn't exist anymore. The power they had borrowed from the District was no longer available, and their influence was waning. They blamed Jerry Maxwell. They blamed Sonny Vergara.

Things had changed.

"A lot of people have different views on Jerry," Seibert said. "But he always gave me straight answers. I held him in very high regard. He took so many truly personal shots to hold this fragile community together. He was amazing."

Seibert pointed out that everyone had a constituency, somewhere to go for help or support, except Maxwell. "It's a lonely and thankless job he had." Gene Heath, Authority general manager (1978–90), agreed: "It is a very lonely job."

Chapter 17
Charting a New Course

In the late spring and early summer of 1997, Roy had me prepare a briefing on the Partnership Plan. In the spring, I prepared communication goals and identified the business community as a key interest group. In my years at the District, I had come to appreciate the way business can influence politics. We needed a new kind of outside pressure, and we needed to make sure that we had supporters. Having dodged the involuntary taking of ad valorem dollars by the Authority, the District wanted to control its program to the extent necessary. There was public talk of the District providing financial support for new water projects, but talk about wellfield-pumping cutbacks was limited. My job was to make our plan the plan for water supply, the plan of choice.

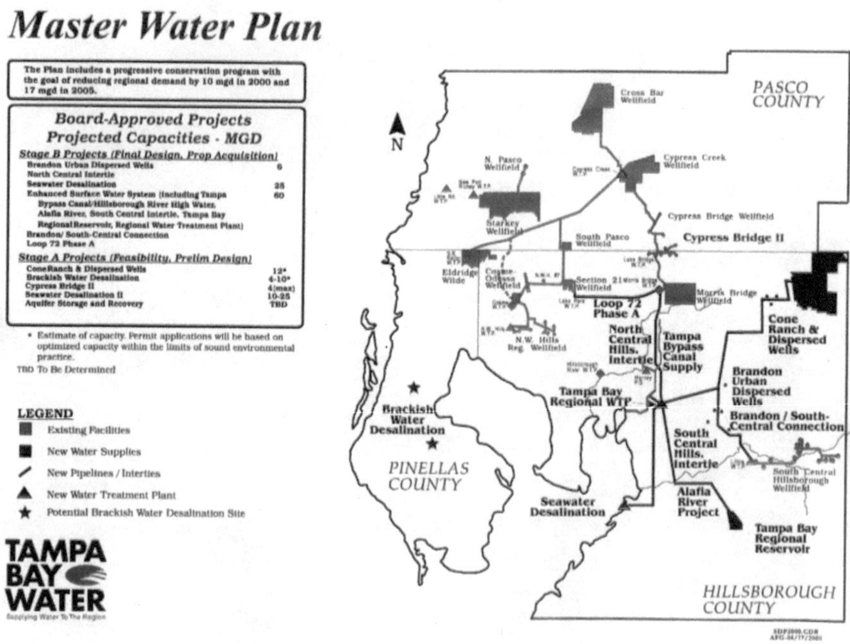

We needed to get the environmentalists on board along with the business community and the media. I gave more than one hundred presentations in four months. Some of them were to key individuals, others in large public meetings. I spoke to virtually every business group from Hernando County (where District headquarters was located) to Manatee County (south of Tampa) and from Pinellas County on the Gulf Coast to Polk County east of Tampa. I stayed on the road, selling the plan with enthusiasm. I had in my package an article from the *Tampa Tribune* about that Korean chip factory considering relocation to Florida. It was looking in the Tampa Bay region. What it needed was clear, fresh, unused groundwater, the very commodity we were fighting over. The availability of water supply was the critical factor in siting the plant. The battle over water in Tampa Bay was affecting the chances of getting the plant and the high-paying jobs it brought. The article and the message of cooperation and partnership were powerful persuasion for the business community.

When I met with high-profile business leader and real estate broker Lee Arnold, he said I had no idea the effect of the water wars on business. I said, "Business goes south," a little too flippantly. "No," he said, "business goes elsewhere." He wasn't smiling. The chip factory went elsewhere.

***

There were things going on in the broader community, according to Steve Seibert. "Suddenly, we heard from CEOs. We were starting to get that 'Listen, this is important' message. The business community was a critical group," Seibert said, not knowing about the District's presentations to business leaders and business groups. "They were a way, way big player. They exercised an enormous amount of influence. And it was good. It was a good policy, and it was good government." Before that, no one had made an organized effort to engage business.

"Those presentations were like an inoculation," Vergara said enthusiastically. "It was like contaminating the whole system with our idea. Business could see the sense of it, and they responded and supported the Partnership Plan. That added pressure on the Authority members to drive them back to the negotiating table." At the time, Sonny didn't realize how well it was working. It's only in hindsight that the presentations made sense. Vergara embraced my strategy when it was evidently working. "At the time I couldn't tell if anything was working," he admitted. "It was so contentious."

On August 8, 1997, Seibert wrote another memo to the Pinellas commissioners. It is clear from the memo that the Authority was closing in on a specific reorganization agreement. His comments addressed budgetary and cost concerns that were raised by Pinellas utilities director Pick Talley and water lawyer Ed de la Parte. Overall, the transition from the current way of being in the Authority to a future way was a deliberate one. Seibert continued to help move the Pinellas Commission forward along a diligent course of change.

In the spring of 1997, Ed Turanchik, Roy Harrell, Sonny Vergara, and Jerry Maxwell met at OysterCatchers restaurant in the Hyatt Hotel overlooking Tampa Bay. They were there to talk specifics and strategy. Each man at the meeting thought he had set the other side straight about how they needed to proceed.

"We kind of mutually weighed where we thought we needed to be," Turanchik concluded. "There was a lot of consensus on it."

By this time Ed Turanchik and to a lesser extent Jerry Maxwell were working out a new governance agreement with the member governments to reorganize the Authority. Direction from the legislature was clear: "Solve your problems, or we'll solve them for you." The people involved decided they wanted to solve their own problems.

The District was trying to support the reorganization but at the same time wanted to get the Partnership Plan on the table, to make it part of the equation. Response to this was mixed. Some people believed the District's money made the reorganization possible. Others found the offer little more than an annoyance, a distraction in the middle of contentious governance negotiations. The District offered $183 million for new water-supply projects and some pipelines to connect them. It offered an additional $9 million per year, for ten years, for conservation projects. The District offered to underwrite 90 percent of the costs for seawater desalination.

Seawater desalination was the one thing the Governing Board would not negotiate. The board was convinced that desalination was the one water source that would guarantee an ongoing, high-quality water supply. District deputy executive Eugene Schiller was assigned the task of working out the details of the agreement and funding to make the cooperative development of a seawater-desalination project possible.

"The bottom line was a fair amount of agreement and collaboration between the chairmen and the executives, and that was real important," Turanchik said.

At least in public the chairmen appeared to be in agreement. Sonny Vergara and Jerry Maxwell met privately several times to discuss the particulars of the Partnership Plan. Toward the last quarter of 1997 it was essential that some of the details be tied down. They met in a room at the Authority offices. Both men remembered the big whiteboard on the wall.

"He constantly asked me to stay away from saying 'cutbacks' publicly," Sonny said. "We tried to accommodate." Publicly and otherwise, Vergara referred to "rotational capacity" or "system capacity" and stayed away from mentioning reductions for as long as possible. "There was a lot of money on the table, and to draw up an agreement, we needed to discuss the details."

Maxwell vividly remembers one benchmark meeting between the pair. "Sonny was a little feistier than usual," he said, with a dismissive wave of his hand. "I had so much to deal with, I just couldn't engage. I told him, 'Look, let's settle in and figure this out or let's forget it, and I'll go do something else.'" Sonny settled down, he said, and the two began to discuss what cutbacks might look like. Sonny flatly denied this happened.

However, with a smile, Sonny admitted that there was far more than one meeting where he was "feisty." He recalled the meeting where the specifics of pumping cutbacks were discussed. "Jerry kept saying they couldn't, they just couldn't," Sonny recalled. "So I got a marker and went to the whiteboard, and we just worked the cutbacks out."

At the end of the meeting, they shook hands and parted, a gentlemen's agreement in place. But after Sonny left, Maxwell went back into that room and took a picture of the whiteboard. He wanted to remember exactly what was on it. He said he couldn't afford any mistakes.

"Several weeks later, we were in a discussion over it, and Sonny didn't recall it right. What he was saying wasn't the deal we made," Maxwell said. "I got the picture and showed it to him."

But Sonny insisted there was no agreement at that meeting. Jerry was using the chart to say something totally different. "I never reneged on a promise, not once, not ever," Vergara maintained, agitated at the insinuation. "There was some kind of disagreement, but I always lived up to every promise I made. Always."

According to Vergara, it was just another long and difficult meeting in a year of long and difficult meetings. "They were down to the point of saying why they couldn't reduce, and I said, 'Wait a minute; let me draw this out.'" He drew the possible future on the board with new supplies, capacity, demand curves, and so on. "We kept drawing it out until we worked it out. We found a plan that would work. And after I walked out, he took a picture of it."

Sonny remembered being confronted with the photo. "It confirmed everything I'd heard about Jerry," he said, letting the matter drop.

Still, the existence of that photo said as much about relations between the agencies as it did about the men involved. In a larger sense, it said something about the character of those who had to make deals to make water work. Despite the deals they struck together, both Jerry and Sonny had to take their ideas back to their respective boards for review and approval. It wasn't easy to gain agreement.

"We had several closed meetings with the Governing Board," Sonny recalled. "I think of them as the Armageddon meetings."

"We knew how it should work and how it could be done," Vergara said.

He could no longer represent the Partnership Agreement as a District proposal without the Governing Board's approval. He told the Governing Board, "You have three choices: (a) we're going to do the partnership, or (b) you can forget about regulation and let them pump the heck out of whatever they want to, or (c) it's Armageddon. Armageddon is where the District pulls out all the stops and puts $20–$40 million in a litigation fund, you get the best lawyers in the world, do what you've got to do."

"I told them the partnership was the way to go," Vergara said.

Even after negotiations on cutbacks became an open matter, there were other issues to be resolved. There are those who agreed that photo illustrated why Jerry Maxwell was considered an outsider at best and duplicitous at worst. The group was in discussions about whether the cutbacks would be to 20 million or 90 million mgd and whether the figure was going to be based on an annual average basis or something else. The District automatically assumed it would be 20 mgd and based on an annual average. But Pinellas lawyer Ed de la Parte knew Maxwell wasn't thinking that way, so he pressed on, asking question after question. "Eugene Schiller and Sonny Vergara were like 'Back off; you're just trying to make trouble,'" de la Parte remembered. So de la Parte asked, "Jerry, when you said, 'average,' did you mean 'annual' or some other period?" Jerry responded that it would be an "appropriate period."

"I think I want an answer to that," Vergara said, sitting up, taking notice, immediately suspicious.

Under de la Parte's questioning, Maxwell admitted that he was thinking of another period, like a multimonth rolling average or something. "That's what eventually got into the contract, but it's an illustration of how they were just assuming he was thinking the way they were, but he wasn't," de la Parte concluded.

It would be tough for the Authority to meet the reductions on an annual average. Even on a three-year annual average, they could barely make it.

"Ed de la Parte is an incredible lawyer," said Vergara. De la Parte worked for Vergara at the Peace River Manasota Water Supply Authority. So did Doug Manson. "Ed is extremely effective, but he needs to be controlled, and Pinellas never did that. They just sort of turned him loose, and it was a field day for him."

While Ed de la Parte's influence on the Authority's reorganization was profound, according to Vergara, Maxwell was not a central figure at all.

"I don't think he attended 10 percent of the meetings," Vergara said. "Jake Varn was there, I was there, the utility directors at first. Eventually, it became the member governments and the administrators. But not Maxwell; he didn't have that much to do with it." Vergara was emphatic.

He noted that Jake Varn and himself frequently played good cop / bad cop in the negotiations. "There would be days when I'd hit the roof and scream and call people names and really insult 'em, you know." It was all in service of getting the deal done, Vergara said.

Sonny's temper, well known in water-management circles, blows over quickly, like a summer afternoon storm. While his anger receded, the people it was focused on didn't forget—especially when the outburst was in a public meeting or public place, or even the hallways at District offices. People whispered about Sonny's temper. It stood in stark contrast to Jerry Maxwell's cool, controlled, dispassionate calm.

While Vergara and Maxwell had issues with each other, neither man stood in the way of getting a deal done. That was not so for some others. The utility directors, as a group, resisted the reorganization of the Authority and the Partnership Plan for as long as they could. After that, some took their resistance underground.

"The utility directors were difficult to turn around," Vergara said.

"No question about it," Maxwell agreed. "The utility directors were the single largest obstruction to getting the deal done. They didn't want to give up their facilities, they didn't want to turn them over, and who could blame them?"

But turning those facilities over to the Authority was critical to making water work, and Maxwell, yet again, took the brunt of criticism for his focus on that goal. Some people said that he thrived on conflict. They said that when you research his past, you see that when there was no controversy, he created some. Maxwell sees it differently.

"I take on the really tough challenges," he said, "and I get the job done. Some people get mad, and some people get jealous, but it's part of dealing with the challenge." Jerry shrugged off the criticism. It didn't seem to bother him. Like most people, he wanted to be liked, he said. But it was not his highest priority.

While Maxwell and Vergara worked out positions each agency could live with, the Authority member governments tried to fashion an agreement to work together. Each of the member governments came to the negotiation sessions with a list of must-haves. By then, each member government had at least three representatives in the room: the Authority delegate, the county or city manager, and a lawyer. This group, the Group of 18, as they came to be called, spent many long days and nights hashing out an agreement they could take back to their boards. Typically, they all brought inside lawyers, outside lawyers, utility directors, and other consultants as well. Not all staff members were happy with this reorganization. Many preferred the demise of the Authority to its reorganization.

The fact is by the time the Group of 18 began to coalesce, the lawyers were being sidelined. "They were at the table," said Seibert, "but they were not leading the discussions."

"I credit Fred Marquis, the Pinellas administrator, with helping to get the job done," Turanchik said. "When you take it away from Pinellas commissioner Chuck Rainey, when you take it away from Pick Talley and Eddie de la Parte, you have the possibility of a resolution, and Fred was the only guy who could overrule a county commissioner or utility director."

"Fred also had a twenty-year relationship with Pick Talley," Seibert said. "It couldn't have been easy for him. I was trying to convince him that there was another viewpoint beyond the utility. Pick went through some very difficult transitions."

Still, when they established the Group of 18, Turanchik couldn't resist using a little influence to get the job done.

"Everyone thought the group had been randomly selected," he confessed. "Mostly they were, but I arranged to be in Fred's group for visioning because I knew he would be key to developing a real solution. Fred was very important."

It was all the administrators, Seibert said. "These guys solve problems. That's what they do. Those administrators had close to one hundred years of problem-solving, which was just essential to helping us find our way."

For the most part, negotiations moved along without media coverage. With a genuine effort underway to solve the problem, everyone was ready to hold their breath, withhold their judgment, keep quiet, and hope for the best. Besides, there's not much media drama in covering a meeting, unless you know participants are going to shout obscenities at one another, and these participants were not going to do that.

And yet Ed Turanchik was frustrated that Tampa mayor Greco still refused to join the group. Turanchik felt strongly that if Greco and St. Pete mayor David Fischer would engage in the issue, it could be resolved sooner. Greco passed again. In his place, he sent Mike Salmon, a staff member and engineer, to represent Tampa's interests.

"Greco wasn't a player, in that I don't think he came to two meetings," Seibert said. "But he stayed out of the fighting, and that was important, too. He was the most popular man in Tampa Bay, and the fact that he stayed out of it was important."

"Greco knew that he needed to be part of the regional solution," Turanchik said with great assurance. "He's a regional guy, a regional leader, and if we made an agreement without Tampa, he would look like a schmuck."

Tampa's staff resisted being a part of the "new" Authority. Tampa had the Hillsborough River. It had the Morris Bridge wellfield. It was planning a new aquifer-storage-and-recovery project—and going after federal dollars to do it. State law requires the water used for aquifer storage and recovery be highly treated, almost to drinking-water standards, before it's injected into wells in the ground. The aquifer naturally continues to clean the water. There are potential problems, like the loss of water through migration and the commingling of treated and untreated water. The Peace River Manasota Water Supply Authority had a project that got $100 million from the District for an aquifer-storage-and-recovery expansion.

"The Tampa staff would have these battles and discussions with Authority staff and others," Turanchik remembered. "Then, in the end, I would call Mayor Greco and say, 'What are you going to do?' And he would say, 'I'm going to have to do something; we're going to have to participate. Do you realize what a jerk I'd look like if we didn't?' 'Yes,'" Turanchik assured him.

"There was one great moment in the negotiation when I knew that we had become a 'group,'" Seibert said. "While Pasco County commissioner Ed Collins was speaking, Pinellas County administrator Marquis was delivered a note to take an important call. He got up and rushed out of the room. Collins stopped talking and said, 'Did I piss him off?' He was concerned. It was a recognition that we were delicately poised here trying to get something done. I just sat back and laughed. 'We are now a group. We are a team,' I thought to myself."

The Shifting Battlefield

By Labor Day 1997, the District felt like a different place. District staff were working on establishing minimum flows and levels, the Partnership Plan, and many other issues. Since Pinellas had sued to force the District to set minimum levels, staff had been scrambling to fit that work in, along with permitting, planning, and all the other ongoing statutory mandates and supporting the Tampa Bay initiative.

The move by Pinellas was designed to overwhelm the staff. It was a new distraction. A new tactic. It was working. The setting of minimum flows and levels—setting a sustainable yield of withdrawal to avoid harming the water source—was a state requirement. The District had a lot of work to do to meet the statutory-mandate deadlines. De la Parte had sued to force us to set a schedule. That schedule was tight, and the staff worked furiously to meet it.

New rules had to be developed, methodologies had to be set and implemented, and local governments, environmentalists, and others had to be included in the process. It was clear that this would be as contentious as every other attempt to place limits on any water resource. The District regulatory staff established a working group of scientists to collaboratively study the problem. Environmental organizations, citizens, and others were invited to sit in. Tasks were divided into subgroups, each with a member of the District's regulatory staff. Their work proceeded for months while Sonny Vergara, Roy Harrell, and others focused on the Partnership Plan, settlements to the emergency order and water-shortage orders, and the many ongoing responsibilities of the District—which extended far beyond Tampa Bay.

In this new District administration, and this new period of cooperation with the Authority and its member governments, some technical staff felt pressured to provide what they believed were less exacting environmental-protection standards—a phrase I coined so that we'd be ready to explain minimum flows and levels to the public. Environmental protection standard is a much better descriptor than minimum flows and levels. No one from the District really wanted to work on minimum flows at this point. There were more pressing matters. But that was exactly Pinellas's point in forcing the issue. Keep the staff busy, too busy, spread them too thin, wear them out, and wear them down. At least that's what the staff thought was happening. It was a tactic Pinellas had used before. Maybe they'll make mistakes. Before it could begin to negotiate with outsiders, however, the District needed to reach internal agreement. Roy Harrell thought that Pinellas might have been too smart for its own good.

"Never assume it's intention when incompetence could be the cause," he said. "Pinellas was trying to keep staff busy, but it almost backfired on them like the lawsuit against activists or the PR program."

"None of us realized it at the time," Harrell said. "We were so busy trying to solve the problem, that the staff took an extreme position—an indefensible position—on minimum flows and levels. When we finally turned our attention toward it, we had a problem. We had a statutory deadline to meet, and a staff proposal was 'ivory tower' and would have been defeated in court."

The internal fight over minimum flows and levels was like the fight over rainfall and pumping. The same people were on the same sides, too. In the past, these technical fights were sorted out and put into perspective, and a staff position was formulated for presentation to the Governing Board. Under the new District administration, the staff operated in a world where the potential reaction to policy appeared to guide policy development. If a stand was judged too harsh, it might be discarded for something less dramatic and less effective. Science used to guide policy. Staff meetings and briefings were held to negotiate the science. Jake Varn, Vergara's old friend, now working as outside counsel for the District, coordinated some of the efforts. In meetings, he appeared to agree with staff. Occasionally, when Vergara sat in on the meetings, Varn sided with the staff. Roy Harrell had some strong opinions about minimum flows and levels, too, but they were inconsistent with the staff position.

Some of the senior staff begged Harrell not to shove his position down technicians' throats. But Roy was sure that the Partnership Plan would fix the problem. Minimum flows were a sideline, an annoyance, and a requirement. A supply-side solution, developing new supplies, would secure the region's future, he believed. The District just had to get beyond the minimum-flows-and-levels deadline to give it some time.

A Governing Board special meeting was held at the Tampa Service Office on September 8–9, 1997. It was a smaller, older boardroom than the one at the headquarters. It is less comfortable and has less technology. It smelled a little stale some days. The meeting had standing room only. The board had been under tremendous pressure on minimum flows and levels. The Florida Department of Environmental Protection had advised it that it had to act to meet legal timelines. Consequently, what was planned as three one-day meetings over three weeks was now crammed into two full days with action required by close of business the second day.

The first day was filled with staff presentations and public comment. The second day was dominated by presentations by District deputy executive David Moore. It was confusing and hard to follow. This was unusual. Normally, ecologists, hydrologists, wetlands specialists, and others presented. Not this time. Moore did everything—including answering the questions. The Governing Board was under pressure and had to act. Some of the pressure came from those who suggested the District should take the hard line. Others said they needed to back off. All accounts were read into the record.

A representative from the Florida Department of Environmental Protection addressed the board and told the members the department did not concur with the step they were about to take. She said that if the Governing Board accepted the staff's recommendation, Dave Moore's recommendation, the department believed significant harm to wetlands might not be reversible. It would be too hard to change it afterward.

A letter from Senator John Laurent was read into the record. He advised the board to take into consideration other changes beyond pumping. Board member and former chair Joe Davis noted the Florida legislature had already adopted a more permissive standard for significant harm, and Senator Laurent's letter seemed to contradict the Department of Environmental Protection.

During public comment, a long line of activists begged the District to adopt minimum flows and levels that would stop the harm and help the environment recover. Gilliam Clarke, Judy Williams, Eileen Hart—the Ladies of the Lakes—and others implored the Governing Board to stay the course and finish the job it started in 1994, almost four years before.

For months scientists were working toward agreement on a methodology and some levels. Not just District scientists but technicians representing the different local governments, environmental organizations, and activists. The outsiders worked hard to reach agreement and were willing to support the District. But the recommendation made to the Governing Board was not what they had worked on, they told the board. It was something that had not been discussed, reviewed, and debated. There were four alternatives for the board to consider, with 1 being the most restrictive and 4 the most permissive.

"When we finally saw what the recommendations from the committees were, we needed alternatives," Harrell said. "But time was running out. We all wanted cutbacks, but we also didn't want another defeat in court. Alternative 1 was unreasonable. It was cutting permitted quantities from over 150 million gallons per day to less than 40 mgd. It was absurd. Where was replacement water going to come from? And, more than that, the cutbacks in Alternative 1 were not consistent with what the legislature intended. What good would it do to pass something we couldn't enforce?" he asked.

Board member Rebecca Eger, an elected tax collector, from Sarasota moved for the adoption of Alternative 1. After some discussion, board member Ray Campo from Tampa amended the motion to allow the rule to sunset in two years. A sunset provision allows a rule to cease being in effect, automatically, unless acted on to extend or modify it. It was a way to meet the deadline without being trapped in the decision.

The tension in the room was evident when Joe Davis snapped at board member Sally Thompson, who said she was confused. "Well, you'll just have to vote confused," he said.

When the vote was called, all but three board members supported it. Roy Harrell, Joe Davis, and Ron Johnson from Polk County voted against Alternative 1. All three of them are lawyers. They believed the rule didn't meet the legislative intent. They understood that in taking the extreme position, their fellow board members wanted to protect the environment. The District was making a terrible mistake—handing Pinellas the opportunity to begin litigation anew.

"We gave the board members too much, too fast," Harrell said, thinking back on it. "It was too much, and they erred on the side of protecting the environment, which is understandable."

As was typical of him, Harrell took it as a challenge to work harder on a solution. Eventually, the minimum-flows-and-levels cutbacks would be part of the programmed cutbacks from existing sources, phased in, as new supplies were brought online. Roy (the sitting chair), Joe Davis (the former chair), and Ron Johnson (the Finance Committee chair) understood the District's position in terms of the law. Roy and Joe also understood it in terms of politics. To win, you must stay in the game. Alternative 1 took the District out of the game, Roy believed.

Still, anyone who was at the meeting knew how awful it was. All the technical staff were at the table, and yet only Deputy Executive Director Dave Moore made the presentation.

"He didn't have enough time," Harrell said. "I asked Dave to give us some alternatives. The working groups had been going on for months. Dave had no time at all."

Everyone in the room knew that something was wrong. This was not the standard-quality work product of District staff—including Dave. The staff performance was poor, the wrong people were presenting, and the Governing Board was split. This was not the same organization that had been led by Peter Hubbell and Mark Farrell through one challenge after another. This was something else.

It had taken less than a year.

***

A few days later, I received a call from Jean Heller, the *St. Petersburg Times* water reporter. It was clear someone had discussed the events of the meeting with her, and she had questions. On my voice mail, she said that it appeared the Governing Board had voted on something it didn't understand. Coming from the newspaper, the comments were meant as a political weapon likely fired by Pinellas County. This was exactly the storyline that Pinellas wanted to advance. I was afraid that someone was trying to drive a wedge between the staff and the Governing Board.

In the old days, I would have just called the reporter back and answered the question. Then again, in the old days, this wouldn't have happened. On this day, I went upstairs to see Vergara. He got Roy Harrell on the speakerphone. I think there was a lawyer in the room, maybe Dave Moore, too. I explained the question and asked what they wanted to say about it. We kicked around a couple of answers, mostly the kind that blamed staff for the presentation. In truth, Roy had told Dave to make the presentation. He'd given Dave the order shortly before the meeting. There was simply not enough time to prepare.

Unlike the old days, when Mark Farrell, Peter Hubbell, and I would have argued with Roy, there was nobody to do it now. We negotiated the language of my response. I spent too much time building staff morale. I was not going to make any of them look bad. We agreed that I would say, "The staff could have done a better job." In my mind that meant Dave. There was no one present who could say that what they saw was Dave's best work. I went downstairs and called Jean. I gave her the District's answer. Our goal was to answer the question without making anyone look bad.

On September 15, 1997, the *St. Petersburg Times* quoted me as saying, "We feel like we could have done a better job of explaining our position to the board members." It was what I had said, and it was true.

It was the last time I spoke on behalf of the District. I called Michael Molligan, the media coordinator, into my office. "You're going to have to take the media calls," I told him. "I can't do this anymore. If you're told to say something you can't stand, bring me into it. Protect yourself."

Staging Events

The regular meeting of the Governing Board went late on September 23. A handful of people remained after 7 p.m., when it came time for the final comments of the day. Jeff Mathias, an activist who always needed a haircut, stood up. He blasted Dave Moore and me over the minimum-flows-and-levels meeting. Dave was blasted for his presentation. I was blasted for the comments to the *Times*. I'd been in my office packing up for the evening when he started to speak. I turned toward the television sitting on my desk.

"I have some issues to bring up, and they go toward credibility," he said. "I think I saw the worst presentation ever made by District staff. Dave Moore seemed inept and confusing. I want to say that I feel each one of you board members are very capable and competent. There was an article in the *St. Petersburg Times*, and Honey Rand was quoted, and there's a credibility issue here. She does not speak for the environmental community. I'm going to be very skeptical of anything that comes out of the PR department for some time."

"That crazy SOB," I thought. "Someone should slap him." I turned to finish packing as Roy asked the board if it had any more business to discuss. Ray Campo, a small, wiry man who spent most of his board time on regulation and the balance trying to undo Roy's initiatives, spoke up.

"Mr. Chairman," he said. "I wasn't going to bring it up, but I have to. District spokesman Honey Rand indicated by a quote in the newspaper that the board didn't mean what they passed by majority last Tuesday. I hope that I, as one member of this board, don't have to ever read a quote of that nature again. Whatever this board does, we do it with the best conscience and information we've got."

"It's very easy to point fingers at others," Roy interrupted. "But when you point a finger at someone, you have four pointing back at you. It's not going to get us anywhere. And for every person under criticism, they've done many glorious things for this District, whether it is the board or the staff. We're going to have to bite our lip and get through this process for the residents, the environment, and the water supply. Hopefully, we can do something that will address these big issues," he concluded.

Sonny said nothing.

Roy banged his gavel, and the meeting was over. The TV faded to black. I stared at the blank screen, stunned. I was sick. My heart hurt. My stomach flipped. The hours, the days, the nights, the weekends, the holidays, the risks, the rewards all seemed like so much nothing in that moment. Everyone knew I was there. I had to go into the boardroom and say my goodbyes for the evening. I didn't want anyone to think the comments had gotten to me. I took a deep breath, got up from my desk, and moved toward that room with what I was certain must look like assurance. The first person I saw when I walked in was Mark Farrell, who was there for a client.

"Did you see that?" he whispered as he bent toward my ear.

"Mm-hmm," I responded in the affirmative, looking around the room, not wanting to be there.

"Sonny should have said something. Anyone who's been a supervisor knows that. It's basic," Mark said, his hand firmly on my shoulder. "You defend your staff."

People walked by murmuring things like "Did you see that?" and "Don't worry, it's just Ray" and other things that were either meant to fill the gap or provide some comfort. I made eye contact with Roy across the room. He nodded a little smile, a small wave. Mark started to head out the door, and I followed him. I wanted out of that room. As soon as we were in the entryway, he said, "Honey, you didn't deserve that." I started to cry. I was so, so tired; I was so miserable. I didn't love my job anymore, and I was watching everything we'd all worked so hard for unravel. Ray's comments meant nothing—but the fact that they were said meant everything. Not to me personally, but to the effort we'd made. Mark and I took five quick steps out the front door into the heavy humid air.

"You've gotta get outta here," he said. "This is bad for you."

That night I felt every burden for the organization, for the activists, for those who would stay. But I couldn't do it anymore—not even for Roy.

I sobbed. I gasped for air. Not for myself but for what we were losing. What we'd lost.

In all we'd been through, Mark had never seen me lose it. It came as a shock to us both. I was very tired. We walked around the corner, and he said, "Quit now," which was funny because it was Mark who'd been helping me keep things in perspective to stay, to stay to help Roy.

Whenever I wanted to quit, Mark would remind me of people who had walked out without a plan. None had done well. I was too visible, he told me, too high up in the organization. "What you need is an exit plan," he said. "It doesn't matter what it is, but it should be multiple steps so that it takes a while to implement. When you're ready to quit, don't quit on the spot [before] you implement the plan."

My plan included proofreading my résumé, updating my portfolio, writing the letter, and a few other things. All those materials were at home, so implementing the plan was delayed at least twelve hours—time to cool down, get refocused, and do what had to be done. When it was time to implement the plan for real, I would know it, Mark had said.

For a moment we stood there, shuffling our feet, when a red truck came around the corner too fast.

"Oh man. Ray Campo," Mark said.

I tried to walk away. I could hear Campo hollering behind me, "Come here, come here."

"No," I said. I got two steps away when Mark grabbed my arm.

"You have to talk to him."

I spun toward him. "What do you want, Ray," I said, dispensing with the formalities.

"You don't understand, Honey," he said.

"I didn't deserve that."

"I was trying to show them what they hired," he told me. "He didn't say anything." He meant Sonny.

"You didn't have to use me to do it," I said. Campo was hanging out of the window of his shiny red truck. He tried to grab my hand. I stepped back. I started crying again. Maybe I hadn't stopped.

"I have to go," I said to Mark, and I left the two of them standing there.

The next morning Gilliam Clarke called to tell me Ray was concerned about me.

"He was trying to make a point about Sonny," she said.

"Gilliam," I said, "He was wrong. Period. What good did it do him?"

It did Ray no good, but it had a profound effect on the senior staff. The message to staff was clear. No sacrifice is too great. No matter how much you care, no matter how much you put yourself on the line, you are expendable. Staff were expendable. It is interesting that Dave Moore and I took the same bullet. We represented different approaches to this new administration, and yet we had the same outcome: if you go out on a limb, you are on your own. Dave had done everything he was asked to do, even when it was directly counter to what he had done previously. I tried to find a way to stay centered on my beliefs and to help Harrell get the partnership in place.

Somewhere along the line, Sonny wrote a letter of apology to Dave for not defending him. It was distributed to all staff members. There was no apology for me. A District department director once told me he kept his copy of the letter in a file, a constant reminder of the cost of being visible.

When I resigned in October 1997, everyone presumed it was because of that meeting. It wasn't. I just didn't want to be there anymore. I turned my resignation in to Ed Helvenston, the general counsel after I called Roy and told the public-communications staff. I felt relieved. By the time I left in January 1998, Roy would have less than sixty days left as chair. Then he would be gone. I had done what I promised. I stayed for Roy, and I helped him to put the Partnership Plan in place.

"In 1997–1998, there was a kind of weariness, a change in attitude at the District and with Pinellas and St. Petersburg," water lawyer de la Parte said. "It was an opportune time to work out the issues. That's why the partnership was great. It gave the money for new projects and got significant cutbacks as part of the deal. Most people might generally agree that the Partnership Plan worked well. If you ask them about the reorganization, they might not think it was so great."

In the interim, Sonny Vergara selected a new assistant executive director, Gene Heath, the former general manager of the Authority.

"Two years before Peter Hubbell decided to leave, I used to joke about it with Gene Heath," Sonny said. "'You and I ought to go up there and straighten that place out,' I'd say, and he'd say, 'Sure, no problem. You go get the job, and I'll be your assistant.'"

Sonny was amused at the way things turned out. "We joked around a lot, and then when I got the interview, and it got serious, I called him and asked if he was still interested. He needed to think about it, but his résumé matched perfectly with what we needed."

"The District had to change," said Roy Harrell. "The Partnership Plan and the minimum-flows-and-levels deliberations were indicative of what was coming. I'm not sure what was right or wrong."

Change was coming—Roy could see it and helped lead the way.

Some of the District staff worked with Gene Heath years before when he was with the Authority. "He's a nice guy," they said. "The least contentious of all the Authority general managers." Very high praise, indeed.

No words, no single symbolic gesture could have been more expressive. It was a new day, or maybe an old one, at the District. The cycle continued. It was an El Niño year. It was raining. And it was time to go.

Chapter 18
Making the Deal to Make Water Work

The Governing Board and the Authority board met on January 9, 1998, for what was advanced as a watershed in intergovernmental relations. The two organizations had met face to face fewer than a handful of times in their thirty-year history.

"We had this meeting with both boards together," Pinellas commissioner Seibert recalled. "Roy Harrell was going on and on about the politics and the science and this and that, and Pasco County commissioner Ed Collins interrupted and deadpanned, 'Yeah. It's political science.' And everyone just cracked up, humor relieving the tension."

But arguably, the most valuable tool of all was the money. Opinions are mixed about whether a deal could have been made without it.

"It was Roy's idea for that pot of money that kept everyone at the table negotiating," said water lawyer Doug Manson. "Without it, the Authority could never have come up with the funding to buy member facilities."

The District's money—the tax dollars allocated to the partnership—ensured that the customers wouldn't have sticker shock. The cost of water would increase over time, but so does everything else.

"It was as much about the money as it was about the District's willingness to offer the money," Manson concluded.

"Roy was there pushing the money, and it was a big deal," said Seibert.

Ed Turanchik disagreed. "In the end, it made the economics of the situation much better, but we had the reorganization done before the Partnership Plan, so it wasn't critical."

But Manson was right. Everyone in the Group of 18 knew the money was there. Of the eighteen, at least six, the county administrators and the city managers, would be responsible for coming up with the money to fund reorganization.

"Their reorganization negotiations were stalled," Sonny Vergara said emphatically, "until we offered the money."

Local Sources First: Good to the Last Drop

Gilliam Clarke had been ill in the months leading up to the 1998 legislative session. Lung cancer had forced an operation that her doctor believed would be sufficient to stop the disease. They caught it early. She was weak but feisty as ever when she went to the legislative-committee meeting for the final chance to make Local Sources First law.

Senator Jack Latvala was chairing the Senate Natural Resources Committee meeting, as usual. Gilliam's husband and water-activist partner, Silbourne, had had a recent cataract operation that made it impossible for him to drive, so Gilliam did. She had never been so tired in her life, she said.

She walked into the basement of the building in Tallahassee. It had that old, musty smell that government buildings often do. "Two staffers came running up to me and gave me a big hug," Gilliam said, tears edging close to the memory.

Gilliam walked into the meeting room and sat down. As she did, she heard over the speaker "Missy, what are you doing here?" It was Latvala. He continued, "I'm not taking public comment." Gilliam wept, remembering that day, remembering that moment, remembering how tired she was. She went to the microphone, and she told those senators they needed to support the bill. It didn't apply to the rest of the state; it applied to Tampa Bay, and Tampa Bay needed it.

She had one last testy exchange with Senator Charlie Bronson, former Florida agriculture secretary, who wanted it to cover the state. Gilliam was afraid the move would hurt support from the Miami/Broward delegation. The vote was called.

In 1998, one year after Mark Farrell and Peter Hubbell left the District, a few months after Roy Harrell and I left, and just before Joe Davis departed, Local Sources First became law. It was not everything we'd hoped for, but it hadn't been totally changed, like when you go from a book to a screenplay. You could still see the original in there.

"I just bawled," Gilliam said, wiping her eyes. "It was Jack's courtesy that time. It still gets to me. You know, Jack is rude, and he is crude, and he looks exactly like his mother, so he's also kind of cute, but he has the courage of his convictions if you get past that tough exterior."

Anger Evaporates

In the spring of 1998, I taught an environmental-communication class at the University of South Florida. I invited two speakers: water activist Judy Williams, and St. Petersburg utilities director Bill Johnson. I knew the students would like Bill. From him, I wanted them to learn that there is another side to the story, that good people can disagree, and that those disagreements can have costly consequences. From Judy, I wanted them to learn about the transience of power. Bill gave the presentation I expected. He was likable and sympathetic. The students listened politely and heard him call me the "assassin" in the District's communication efforts during the water wars.

The lesson they learned from Judy was harder. Judy was still fighting the specifics of the water wars. She stood at the whiteboard, talking about geology and the technology of alternative water supplies. The discussion was too complex; the students were mostly unprepared for it. I went to the board, and while she was talking, I drew a table. The table listed years across the top and organizations and groups in the left-hand column. It listed the Authority, the District, the activists, Pinellas County, Pasco County, and a couple of others. When she took a breath, I interrupted.

"Who has power in the water issue now, Judy?" I asked. She hesitated.

"Who has power?" I pressed her.

She didn't want to say the words.

"OK," I said. "In 1993, who had power?"

"Pinellas County," she responded tartly. She didn't like this exercise.

"Anyone else?"

"The Authority."

"In 1994, who had power?"

"The activists." It had been the year of Steve Monsees, the year the District awakened. It was the year Judy joined the activists. She smiled.

"Anyone else?"

"The District."

"In 1995, who had power?"

"The District," she said, a big grin spreading across her face. It was a good year for us all.

"In 1996?"

"The District," she responded. Another good year.

"In 1997, who had power?" She hesitated.

"When did Mark and Pete leave?" she asked.

"January 1997."

She frowned.

"Who had power in 1997?" I pressed her.

Judy pursed her lips.

"Who?" I insisted.

"The Authority." Her voice was flat, deflated.

"Who has power now?" I asked. I wanted the students to get this lesson. I wanted them to understand it.

Judy wouldn't answer.

"Who?" I demanded. It was an order. We'd fought the good fight together.

Tears came to her eyes. "Jerry Maxwell." The activists, particularly the Ladies of the Lakes, hated Maxwell. They thought he had his own agenda, they were sure it didn't match theirs, and they were frustrated with their inability to make things happen as they had before.

I asked my students, "You can make a difference, but can you make a lasting difference?" I asked Judy, "Did we?" By then, I wasn't sure anymore.

"They didn't understand that with minimum flows and levels and the Partnership Plan, we achieved everything they wanted," Roy said.

There was a disconnect between the District and the activists. They saw the same outcome, but to each it meant something different.

Closing the Deal

Bill Johnson liked to tease me that the water wars started a month after I got to the District and ended six months after I left. While I continued to insist it was coincidence, it was funny.

Since late 1996, the Group of 18 met more than twenty times as they hashed out the terms of the new governance agreement. The Authority found some whimsy in it and provided T-shirts to them. The front says "I survived the West Coast Regional Water Supply Authority Group of 18." The back says "Building Blocks for Regional Solution: 1,320 cans of soda, 624 bagels, 480 deli sandwiches, 432 donuts, 48 pints of Chinese food, 44 pizzas, 18 pounds of lasagna, 2 spiral cut hams, 1 birthday cake."

In all those meetings, the public never came out, according to Mayor Fischer. "People expect to get water. They expect to turn on their taps and have it there. They don't care where it comes from, only how much it costs and how it tastes."

The Partnership Agreement and the reorganization agreement were signed in a ceremony at Florida Aquarium at the Tampa Bayfront on June 10, 1998. The Authority gave out crystal water glasses. I brought one back for my office. Joe Davis and Roy Harrell gave me theirs as well. I was present when the deal was signed, but it wasn't my deal anymore. I had moved on. Still, I have the glasses, and they're mine, just like the experience of the water wars, of developing a better water policy—or forcing someone else to.

Amazingly, there were few casualties in the water wars. Most of the veterans were there at the signing: Doug Manson, Ed de la Parte, Pick Talley, Peter Hubbell, Mark Farrell, the Governing Board, the Authority board, staff and consultants—everyone but Chuck Rainey.

Ed de la Parte believed that anything short of the partnership would have left local governments dragging their feet to meet District mandates, fighting them every step of the way. The partnership made it possible to come to the table willingly and to stay there.

With the signatures of elected officials representing the entire Tampa Bay region and several environmental agencies, the old West Coast Regional Water Supply Authority was gone. In its place was Tampa Bay Water, the regional water wholesaler, among the largest utilities in Florida, and certainly the District's largest permittee.

"They didn't really have to change the name," Doug observed, "but the old name had too much baggage."

The question was, would people give the "new" agency with the same players a fresh start? The answer was mixed. From a functional and political perspective, the Authority was gone, and Tampa Bay Water was new. From the public's perspective, little had changed. The same people were in place. They had the same values. Many had the same jobs. Could they change the way they did business? Did they need to? I called this *the same old girl in a new dress*.

Still, I believed that the Partnership Plan and governance agreement were the region's best chance for peace.

"We had a situation where, into perpetuity, Pinellas County and St. Petersburg would help subsidize the cost of developing new water supplies for growth," said Ed Turanchik. "That's in exchange for us in the short term sharing the cost of replacing the water. Any reasonably bright county commissioner today would think that ensuring water supplies in perpetuity is a good deal. Pinellas and St. Petersburg will pay 40–60 percent of the cost of new supplies."

Since Pinellas's and St. Petersburg's growth was fueled by the water from and environmental damage to Pasco and Hillsborough Counties, it seemed fair. And, apparently to those who signed the agreement, it was a reasonable price for peace.

"It was a remarkable thing," Turanchik said about the historic agreement. "I think it's the best piece of environmental legislation this area has or will see for ten to fifteen years. It's going to have lasting benefits."

"I wish I had proposed it ten years earlier," Roy Harrell said.

But was the conflict over?

Mark Farrell and Peter Hubbell were doing well in their consulting business. For a time, they'd represented Pasco County. Through 1998 and 1999, they worked for Hillsborough County. All of Tampa Bay Water's new water-supply projects were going to be in Hillsborough County. To ensure their future was protected, Hillsborough set aside more than $10 million to fund a team of consultants to watch the permitting process and protect its interests. Mark and Pete's expertise and their relationship with District staff should help ensure long-term protection of the resource. Among those first difficult tasks was the completion of minimum flows and levels—the amount of water that could be taken from the aquifer and from rivers. By 1998 Pinellas, Tampa Bay

Water, and the District were largely in agreement over new rules for minimum flows. Hillsborough County and Tampa were not very happy, but for different reasons.

Pasco County was out of the fight. Having ensured that the first wellfield cutbacks would occur there, they were satisfied. Pasco had been the victim in this drama since the 1970s. No one wanted to target Pasco for new water-supply development, so it watched from the sidelines. Until now. In 2023 Tampa Bay Water short-listed new water supplies, and on the list was groundwater from northeastern Pasco County. Water wars were fought over wellfields in central Pasco. This is a different world, and nearly all the water warriors are dead or moved.

Tampa, in contrast, was singularly unhappy. After signing the reorganization agreement, it was now faced with giving up the Morris Bridge wellfield. Kathy Fry (the Tampa staff attorney) and Mike Salmon (who sat on the Authority for Tampa) wanted to make sure the Hillsborough River was secure. They didn't trust Sonny Vergara. For his part, Vergara had decided, like Turanchik, that appealing directly to Tampa mayor Greco for cooperation would be more productive. Greco was willing to meet with him but wanted his staff to be present.

A meeting was arranged with Sonny Vergara, Mayor Greco, Mike Salmon, and Kathy Fry. The minimum flows and levels under consideration might reduce the amount of water available to Tampa from the Hillsborough River.

To complete agreements on minimum flows and levels, Tampa needed assurances. Sonny wanted as little protest as possible and believed it was better to negotiate than to litigate. To do that, Tampa needed a guarantee that it would keep the quantities it had already been allocated. The Morris Bridge wellfield was used by Tampa during low-rainfall periods to augment river water. Giving it up as part of the reorganization meant that Tampa needed to keep every drop of the river and needed to be sure that the District would protect its interests in negotiations over minimum flows and levels—a process closely monitored by environmental groups.

Vergara was trying to persuade the mayor that joining Tampa Bay Water in its reorganization was best for everyone concerned. It was a powerful argument. Mayor Greco is a regionalist at heart, and water, though it was an issue, wasn't where he wanted to spend political capital.

In the meeting, Sonny appeared to promise the mayor that Tampa would be taken care of—something the Tampa staff didn't believe or trust.

"So, what you're saying," said Mike Salmon, "is that you're going to take care of Tampa's interests." It was a statement, not a question.

"No," said Sonny, surprised by the assertion. "Our job is to take care of the environment, not Tampa's interests." The staff had made their point, and the mayor, who seemed only marginally interested until then, sat up.

He leaned forward and summarized the conversation in three or four succinct points, demonstrating that he completely understood everything that had just happened. He pointed his finger toward Sonny Vergara and said, "You're not going to screw with my river." The meeting adjourned.

Vergara didn't remember the meeting. But he said there were meetings with Greco and the staff to give Greco some information he might not have and might need in order to move beyond staff recommendations. Sonny recalled that the Tampa staff were not happy. Still, he felt he made a connection with the mayor.

He didn't like going around the staff, but in his mind, there was no other way. With the staff and the mayor present, Sonny felt he could make his case, and they could make theirs.

"By then, I was not really into confrontation," he said. "I used to do things like former District executive director Pete Hubbell, but I found out the hard way that when you stand up and give people what for, it has to serve a very strategic purpose. If I wanted someone to do something, I didn't bother to go tell them; I did everything around them so that they did it anyway. And that's the way to get it done." That's what he was trying to do with the mayor.

"Sonny was the right man for the climate," Roy Harrell said.

The relationship between the chairman and the executive director is typically close. Together, they shepherd the agency through the tough times and around the political land mines.

"We didn't need to be in their face anymore," Harrell said. "We needed to finish the job. Sonny helped do that."

Exit Stage Left

With the signing of the governance agreement and Partnership Plan, the region had a plan, the money, and the political will to make water work. But in a strange twist of fate and a little political hardball, all but two members of the nine-member Tampa Bay Water board were replaced. While not devastating to the program, the abrupt change was disruptive.

"That's the X factor in the whole thing," Steve Seibert said. "Who would have thought that all of us would leave at about the same time? What were the chances of that?" It's all too clear that the departure of three key members of the Group of 18 had a direct impact on the implementation of the master water plan. "It was the saddest thing because the historical knowledge was gone," said Seibert.

And yet there was potential as well.

"Personality changes helped make the deal possible," Vergara said. "Chuck Rainey leaving 1996, Steve Seibert coming 1996, Mark and Pete leaving 1996, me coming 1997, Roy leaving 1998. All those personality changes were key."

Three other critical personnel changes occurred in 1998: Pasco commissioner Ed Collins was defeated in the November election, Hillsborough commissioner Ed Turanchik resigned to become president of the Florida 2012 initiative—Tampa Bay's effort to get the Olympic Games held in Central Florida—and Pinellas commissioner Steve Seibert was chosen by Governor Jeb Bush to lead Florida's Department of Community Affairs. Their collective departure within thirty days left a huge void of corporate knowledge at Tampa Bay Water. Some of the new appointees to Tampa Bay Water's board were first-time elected officials, fresh from the campaign trail.

After Ed Turanchik's departure, environmentalists and the newly elected Hillsborough County board members claimed to despise the new regional agreement. They did so vehemently and publicly. They distrusted Jerry Maxwell, they disapproved of the new water projects, and they thought that Ed Turanchik had given away the farm to make the deal. Turanchik, who was known as a committed environmentalist and who worked diligently to create a regional solution, was vilified by the newcomers.

"The truth of the matter was that Ed Turanchik helped get all of us to pay for projects that would help Hillsborough County," St. Petersburg mayor Fischer said. "We're paying for a reservoir, we're paying for the desalination plant, we've tossed all of our assets into a common pool, and the new infrastructure is going in Hillsborough. He made sure that Hillsborough had more influence on the projects than anyone else. Hillsborough ought to feel pretty good. We're paying for their growth."

"People who don't want projects in Hillsborough County are the worst kind of NIMBYs," said Turanchik. A NIMBY is someone who says, "Not in my backyard." They might not oppose a project unless it's built near them. "You're not supposed to oppose a project," Turanchik said, "just because it's in your backyard. You're supposed to support the projects with minimal environmental disruption. When you compare potential impacts of the alternative projects to the real impacts of the traditional projects. . ." He didn't finish. He was disgusted by the narrow-mindedness, the self-centeredness of the NIMBYs.

"They so didn't get it," Manson said, shaking his head in disapproval. "Ed Turanchik had a strange negotiating style . . . almost criminal." He laughed. "It was a wear-you-down style. He would basically exhaust you on an issue, then wait three or four weeks, bring it up again, and cut a better deal."

Hillsborough County came out of the reorganization much better off than it was in the beginning, most observers said. "Turanchik got huge concessions," Manson said, "in addition to getting St. Petersburg and Pinellas to pay for growth."

"It didn't bother me," Turanchik said of the criticism from the new commissioners, "because I come from a position of positive certainty that the deal we cut was the right thing to do."

A colleague of Steve Seibert said it best. "When entering the room, leave your ego and your logo at the door." In truth, said Seibert, "We never really left our logos." Pasco always represented its interests and Pinellas its. It's just that somewhere along the line those interests converged.

"When we first began negotiating the water-shortage orders and the emergency order, we were operating under the old rules, in the old game, with the old constraints," said Seibert. "This thing we did, this agreement we made, was something totally different. There is no obstacle too great that people can't change—if they put enough work into it. What system do you know that works well for twenty years?"

But the world changes around a system, and the people inside the system change, too.

When new board members were added, Pasco commissioner Ann Hildebrand was reappointed to the board. Despite her two-year absence during the reorganization, she had more direct experience than any other board member.

"One of the biggest challenges was turnover in the board," she said. "It was such a technical—such an involved—issue, it took a lot of time to understand it. Sometimes a person barely gets up to speed, and they say adios, and you've got to feel sorry for the staff having to start all over every time."

Mayor Fischer agreed. "When new people come on to the Tampa Bay Water board, it takes time for them to get up to speed," he said. "Sometimes I think because they're spending taxpayers' dollars they should be elected.

But it's hard for elected officials because on the Authority board we must make decisions based on scientific fact, not politics. That's a hard thing to balance, and every year, the representative can change. When the people change, it's like 'Go. Stop. Go. Stop.' Once the water flows, this will become a very boring board to sit on." He smiled, imagining that day about eighteen months away, he believed. "I tell Jerry, 'Stay the course; it's just eighteen months.' But it's a tough job." Fischer worked with Jerry Maxwell every day. He read the papers.

He knows how hard it is.

Pasco commissioner Ann Hildebrand agreed. "But I maintain that he [Maxwell] was the guy for the job where we were," she said.

Looking to the future, she believed the best course for Pasco was to work the plan, choose new sources, and develop them. Only then could the pumping in Pasco be reduced and the environment restored. She credited Jerry Maxwell with the advances.

"He took a very aggressive approach," she said.

Pasco commissioner Ed Collins served for a few months with Ann Hildebrand, but in the November 1998 election, he was targeted for defeat by Gilliam Clarke. Gilliam claimed she never liked Ed Collins and had just played to his ego. Collins noted that he, Gilliam, and the other activists worked closely together from 1996 to 1998. She used to call him El Tigre (the Tiger) for his aggressive approach to protecting Pasco.

Gilliam said she targeted Ed because he was dishonest and he voted with Pinellas County and not for Pasco. She released information to the *St. Petersburg Times* just days before the election that brought Collins's ethics into question. He said that Gilliam wanted the first seawater-desalination plant to be in Pasco County and that his vote for a project elsewhere brought her wrath. The bottom line is that Collins won the absentee-ballot vote—the voters who hadn't been exposed to the negative newspaper articles—but lost the election. Months later, the Florida Ethics Commission completely cleared him.

Tensions Rise

Anyone who thinks this situation and these jobs are not stressful should spend six months trying to follow those developing policies, projects, and politics. The job is hard enough when an agency is under pressure from outside. When there is a feeling that your authority is being undermined, or that someone is intentionally adding aggravation, tempers can and do flare. Typically, these blowups do not result in physical contact; the angry tempers are kept behind closed doors. But the drama of it, when made public, is newsworthy.

One morning the Tampa Bay newspapers ran headlines about Tampa Bay Water general manager Jerry Maxwell trying to choke General Counsel Don Conn. The police were called. Twice. The first time no report was filed, according to Maxwell. The second time it was.

It is difficult for most people to understand the volume and complexity of materials that are produced to support decisions on water. There are technical and progress reports, budgets, memoranda, contracts, and myriad other documents, illustrations, maps, charts, complaints, notes, and proposals. There is no question the staff at Tampa Bay Water were overworked. Many of them were in positions causing them to stretch professionally—not because of inadequacy but because the volume of work was overwhelming. The same is true of board members who have other full-time responsibilities. Given the volume of correspondence overall, some of Tampa Bay Water's board members had suggested that last-minute correspondence was meant to overwhelm them, to hide things, to get them to overlook the details. Board members had publicly and privately criticized Maxwell, and he had vowed to gain control of the problem.

On a Monday morning before a Tuesday board meeting, he found additional materials had, yet again, been sent. Maxwell had made it clear to his staff that this was not to be done. The new board was uncertain enough; there was no need to add aggravation. When he looked into the matter, he found that despite his direction, Don Conn, the general counsel, had faxed the additional information.

"The guy did it all the time," Maxwell told the *St. Petersburg Times*.

Maxwell was furious. He went to confront Conn. In his hand, he had a copy of the fax. Don stood up as Jerry walked in. Uncharacteristically, Maxwell was yelling. "You dumb son of a bitch, what the hell were you thinking?"

Conn responded. In the angry exchange, Maxwell said he reached toward the lawyer to grab his shirt pocket and shove the wadded fax into it. When he did, Conn recoiled as if he'd been struck. Missing the pocket, Maxwell said he caught Don's tie for a second, let go, and left the room.

Back in his office thinking about it, Maxwell was unhappy with his own conduct. He went to Conn and apologized, then returned to his office. Eventually, the police showed up. Conn claimed that Maxwell had his hands around his throat, choking him. The police took statements from everyone and filed a report. Not too surprisingly, it ended up in the newspapers.

On September 24, 1999, the *St. Petersburg Times* reported that Conn's police statement had Maxwell saying, "If you ever do this to me again, you (expletive deleted) I'll kill you." Maxwell was embarrassed. He offered a public apology, but those board members who wanted to, who sided with or sympathized with Conn, were able to once again criticize Maxwell and directed him to attend anger-management classes.

Signs and Symbols

In the spring of 2000, Ed de la Parte was scheduled to depose Mark Farrell on issues relating to minimum flows and levels. Farrell represented Hillsborough County, and Ed de la Parte represented the District. "We weren't paying him," Sonny pointed out. "It supported what we were doing, so we accepted it." In two years, the world had changed. Ed de la Parte was helping the District; Rich McLean (formerly the deputy executive director for regulation at the District) was working for Pinellas County; Mark, Pete, Ed Helvenston (former District general counsel), and Gary Kuhl (former executive director at the District) were all working for Hillsborough County.

A Cold War

"If there's one thing I completely missed," Maxwell said, "it was the politics of having two county commissioners from the same county on the Tampa Bay Water board. Under the old structure, the member could decide, make a vote, and go back to their board to sell it whatever way they wanted. Under this new arrangement, the politics are much more treacherous. One board member can undercut the other—and they do."

"I was the only guy left," Mayor Fischer said, somewhat nostalgically. By summer of 2000 the officials who put together water peace were gone. Pasco commissioner Ann Hildebrand, who was on the Authority board and was chair during the water wars, was off during negotiation for reorganization and back on the Tampa Bay Water board as the chair. Other than what those two knew, the only historical knowledge of what happened in the water wars resided in consultants and lawyers, city managers, county administrators, and utility directors—but not in the hearts and minds of the elected officials that sat on the Authority board. The faces changed, and the politics changed. Even so, ironclad contracts held the organization together. It might not have been the best way, but it was the way.

"Eddie [de la Parte] wrapped things so tightly that even a county commission couldn't undo it," said Seibert. "He used to drive me nuts with his 'If you're going here, we have to make this right.' But he and Doug Manson really made the thing work, from the legal perspective." Seibert pointed out that they were not alone, but it made a difference to him, personally. The contract for reorganization was written by a separate firm, but lawyers representing Pinellas, Tampa, St. Petersburg, and the other governments helped craft the content.

"I felt very comfortable that if Ed de la Parte said we could go ahead, that I knew the liabilities," said Vergara. "Those agreements are as solid as a rock."

De la Parte pointed out that when appointed to the Tampa Bay Water board to represent Hillsborough County, Commissioner Pat Frank said that she would never have signed the reorganization agreement, and she wished it would go away.

"She was honest," Vergara said, clearly more concerned about behind-the-scenes maneuvering than the visible complaining.

Pat Frank was not the only new member who publicly objected to the agreement. Ronda Storms, also from Hillsborough, publicly castigated the Tampa Bay Water staff, berated Jerry Maxwell, and played to a very vocal but narrow constituency. Looking back you can see the seeds of today's national political divides. Following the 1998 elections, the Authority's problems were internal, and frustration was evident.

"They were a little disingenuous," Turanchik said about the Hillsborough commissioners. "The fact is they talked about what a terrible thing this or that project was, but their experts couldn't find any legitimate reason to challenge them."

Even so, Tampa Bay Water's board meetings were preceded by intrigue and punctuated by outbursts from a new group of water activists who didn't want projects built in their communities. Mayor Fischer presided over those meetings with the same calm reason he demonstrated through the worst of the water wars, through the worst of the negotiations, through the angriest attacks from Hillsborough's representatives. A fighter pilot by training, the mayor was unflappable in public. As the chair of Tampa Bay Water, he weathered the public tirades with his usual calm.

"There are too many things to do," he said. "When thunder goes off suddenly, I never jump outside. I go inside. My mind is thinking, calculating, but I don't jump and lose focus; it's that pilot training. 'Ladies and gentlemen, we've lost our wings; they seem to have fallen off,'" he said with a smile. "'But we'll be serving cocktails within the hour.'"

The District occasionally talked tough, publicly holding Tampa Bay Water to the water-supply development and cutback schedules. As of winter of 2001, Tampa Bay Water had exceeded its groundwater-pumping cap before new water supplies could be brought online. Maxwell was still at Tampa

Bay Water, still pushing the projects, still trying to make water work. Some think he was approaching the end of his tenure.

"Jerry was incredibly focused," according to Mayor Fischer. "It was his job to get through that and stay focused on the plan."

"I was amazed he was still there," Turanchik observed.

"Me, too," Jerry agreed, with his charming and disarming grin. "I wanted to stay to get the job done. After that, who knew?"

In the meantime, the activists, the veterans, stayed focused on helping Jerry out—any way they could. They wanted him gone, and sooner was better than later.

"The thinking was 'If I can get Jerry fired, I can blow up the master water plan or governance, and that will delay things,'" de la Parte speculated. "The thinking was that you can affect policy, not directly, but indirectly by demonizing people."

Some wanted to see Tampa Bay Water fail. They were busy every day positioning themselves or their bosses, on the phone coordinating tactics, providing information to activists opposed to projects. They were there, and they were not going away until the water sources they wanted were developed, or the people they wanted were in power, or they owned their own water-supply facilities again.

"Pinellas wanted Tampa Bay Water to succeed," de la Parte said. "If it didn't, there would be a holy war to get their water-supply facilities back."

"We changed the whole world to make this work," Seibert said, hopeful that the plan would survive. "If they were short just a little bit, I hope there's some room. . ." He trailed off, perhaps considering the possibilities.

"I think it's going to work in the long run," Fischer said, "because it was the right thing to do."

In the spring of 2000, Steve Monsees sold his house at a loss of $40,000–$60,000. He moved north, closer to the Withlacoochee River, and assured me that if anyone tried to come and take the water, he'd defend it any way he could.

"I wouldn't be afraid of jail at ninety years old," he said. I believed him. Steve passed away, still angry over the loss and his treatment.

Gilliam Clarke's house continued to sink. In the summer of 2000, her insurance company raised it and poured concrete underneath. Ironically, no one could guess if the cutbacks to wellfield pumping would raise water levels to the point that Gilliam's home would flood. Gilliam and her husband moved to North Carolina, where she drew, painted, and wrote poetry until her death from cancer in 2010.

In August 2000, Betty Tillis was the lead in an ABC nightly-news feature on the desalination plant to be built on Tampa Bay. She was offered as evidence of failed water policy and the turn to a more promising future. Betty told national *ABC News* about her life and how her well went dry, but they weren't interested in local politics. They were focused on desalination as a safe, drought-proof alternative to wellfield pumping. Betty lived off a hose until she moved away in 2001.

Tampa Bay Water's staff were angry with me for giving Betty's name to ABC. They wanted to provide ABC someone else, anyone else, and refused to believe that Betty's problems could illustrate a past gone wrong and a future, corrected course. Betty said that her financial picture had changed, and it was time to think about a new well for her home. She left instead.

Judy Williams moved away, too. Gaye Townsend and Eileen Hart continued to fight for good water policy, but they did so with the scars of earlier battles and the recognition that barring magic, change is slow and deliberate. Eileen Hart returned her Moral Courage Award to the Hillsborough County Commission after it gave an award to "a political hack" in her words. Eileen died of a heart attack in 2021. She stubbornly refused to go to the doctor when the pain began. Roy Harrell died while attending a water conference in Vail in 2006, and John Parker died in a car accident in 2007.

Neil Johnson, the water reporter for the *Tampa Tribune,* said things had changed, and it was hard to get official comment on water-supply development projects in Tampa Bay. Dissenting voices aren't tolerated. Besides, many people were tired of fighting. New water supplies needed to be built. Now, the *Tampa Tribune* is gone, and Neil is retired. Peter Hubbell retired in 2018, and Mark Farrell soon followed.

I spent a lot of time thinking about the people who still work at the District and Tampa Bay Water, and I wondered at their ability at times to sublimate what they really believed whether what happened was the best course for protecting the environment and making water work.

When Rick Scott became the governor of Florida in 2011, he began a deliberate, systematic program that dismantled and disempowered Florida's water-management districts. Executives three layers deep across all districts were released or fled as policies developed over decades were abandoned, disabled, and dismembered. Environmentalists were horrified, business and agriculture were largely delighted, and local governments were concerned. This continued through Scott's two terms. His actions were unsuccessfully challenged until his last year in office, when blue-green algae, red tide, and other foul-smelling, fish-killing maladies plagued wealthy coastal communities and negatively impacted tourism.

In that last year, residents began to share the concerns of environmentalists, and, as the race for Florida's governor developed, candidates were forced to face the carnage. While there were doubts about him, some of Governor Ron DeSantis's first actions were to invest in water-quality improvements and recognize the critical role of science. People were hopeful that he'd begun the process of restoring and respecting the critical role of water resources to the economic engines of Florida. In 2022, he was reelected just as Hurricanes Ian and Nicole ravaged South Florida. His environmental record is inconsistent, and, of course, the rest of the country knows about some of his other egregious policies.

There are new faces at the water-management districts and the Florida Department of Environmental Protection. There are new faces fighting for water and other environmental issues across the state. Some are volunteers and activists; others have sought public office to help lead the way.

Chapter 19
Twenty-Five Years Later

Twenty-five years later, Tampa Bay is better off—at least in its water-supply development. The projects and pipelines cooperatively funded were built and put into place. A recovery program was established for the wetlands and other environmental resources. Despite mostly abundant rainfall, full recovery has not occurred. Big Fish Lake on the Barthle Brothers Ranch has not fully recovered, but it's improved. Steve Monsees's lake is nowhere near the hundred acres it was when he moved there. But things have improved there, too. The tension has dialed down or gone underground. The Authority tries to keep disputes with member governments out of the news and the view of the public. It hasn't always been successful.

"Overall, the District is seeing about a 20 percent decrease in groundwater use—so you're talking 250–300 million gallons per day, roughly, of decline over the last twenty years," said Brian Armstrong, executive director of Southwest Florida Water Management District. "The cutback from Tampa Bay wellfields was 60 or 70 million gallons per day, right there."

He pointed out that while the population has grown, groundwater demand has dramatically reduced. That, he says, is the value of alternative water supplies. Of course, the region was devastated by the Great Recession, but even so, the numbers tell a story. The recovery of the wetlands and lakes—to the extent that they have recovered—is a visible testament to effort and investment.

Water lawyer Doug Manson is quick to point out that conservation—the intentional behavior of residents and businesses—accounts for a 20 percent reduction in regional per capita water use.

The path to improved intergovernmental relations and more water was anything but smooth. Several of the new water supplies and infrastructure projects had significant issues. The desalination plant, developed by a private company, was acquired by Tampa Bay Water before design and construction were complete. The new water-treatment plant produces less water than planned.

The reservoir, a technology used for centuries, had to be taken off-line and rebuilt. After a protracted and bitter lawsuit with the engineering/construction firm, Tampa Bay Water lost, with significant financial consequences for rate payers—customers of the Authority member governments. They had to pay for reservoir repairs along with court and legal fees.

"We knew there would be systemic problems bringing on alternative supplies," said Doug Manson. "There were concerns about the Authority staff's experience in building projects, managing facilities. That turned out to be a valid concern."

The Authority projects planned and funded in part by the District included a desalination plant, water-treatment plant, and reservoir (the District purchased the land). All three experienced significant issues.

"They [Tampa Bay Water] screwed up the water-treatment plant," said Mark Farrell, former assistant executive director of Southwest Florida Water Management District. "They screwed up the reservoir. They built three major projects and screwed each one of them up, costing the customers hundreds of millions of dollars."

Jerry Maxwell was the Tampa Bay Water general manager when the new, alternative supplies were developed. He resigned just before the projects started falling apart. Under his direction, Tampa Bay Water bought the desal plant before it was ready. He was the Authority general manager as the reservoir and the water-treatment plant were constructed.

"We've gone through three general managers," said de la Parte. "We've gone through a turnover of all the board members, and there have been changes in the political bodies as well, so I think Tampa Bay Water has become an institution."

Doug Manson agreed. "What we've found is the institutions are bigger than the personalities."

Still, the Tampa Bay region is better off water-wise.

"It probably helped that the region built the facilities, then the economic downturn slowed growth, stalling demand," said Ed de la Parte. "That probably leaves them with enough capacity to last for decades."

Despite the challenges, the Southwest Florida Water Management District, Tampa Bay Water, and the member governments can all be proud. Significant effort and funding were invested in developing and deploying the new water sources, interconnecting lines, and implementing the revised governance agreement.

So it was perplexing when Matt Jordan, the general manager of Tampa Bay Water, declined to be interviewed for this final chapter. His excuse? He doesn't know anything about what happened "back then." I told his assistant and the public relations person that the focus would be on how well off the region is and the status of the facilities. Unfortunately, he declined. Fortunately, the same week that the interview was requested, he presented a system update to the Governing Board of the Southwest Florida Water Management District. As Eileen Hart said years ago, when you need that bit of information, somehow it just shows up. Learning that I was missing his comments, a helpful District employee got me a recording of his presentation.

More than anything, his refusal to be interviewed suggested that nothing much had changed at Tampa Bay Water. It might have a new general manager, but it still prefers to hide from public scrutiny. Same old girl in a new dress.

Tampa Bay Water's member governments cannot develop water supplies independently. When they signed on to the reorganization, they transferred existing water-supply facilities and future development to the Authority. Only if Tampa Bay Water fails to deliver necessary water can the members develop their own.

According to former District assistant executive director Mark Farrell, "Tampa Bay Water is looking for the next source of water, and the next source of water they want is reuse. What's interesting is that they are trying to negotiate with the member governments to acquire their reuse so they can turn it into drinking water." Farrell is also a former president of the Florida Water Reuse Association.

Farrell was right because in his presentation to the Governing Board, Matt Jordan said, "We're looking at purified reclaimed water to augment other sources. We're working very closely with Hillsborough County on their SHARP [South Hillsborough Area Reclaimed Project] to see what opportunities we may have there. We can work together. We're also working with the city of Tampa on their TAP [Tampa Augmentation Project] program. Reclaimed water is a resource that really needs to be carefully evaluated to see what other benefits can be achieved from that."

By 2018 a new dispute threatened a new kind of water war—over reuse. It was not the first time. A few years back, Tampa Bay Water went to the Florida legislature attempting to have reclaimed water labeled a water resource, which would transfer control of it from the member governments to the Authority. The governments argued that wastewater returned to their facilities and cleaned at the expense of their customers, their users, belonged to them.

The city of Tampa was taking steps to ensure that Tampa Bay Water couldn't be positioned to take its valuable reclaimed water—sixty million gallons per day. A *Tampa Bay Times* (formerly *St. Petersburg Times*) article on the dispute focused on one possible outcome: the demise of Tampa Bay Water. Writer Craig Pittman led with "For 20 years, the regional water supply agency known as Tampa Bay Water has provided the state with an example of how local governments can cooperate to provide a vital resource to a vast metro region. A pair of bills in the legislature could have dismantled the whole thing—and the City of Tampa was backing them."

In Mark Farrell's view, "Why would local governments give treated wastewater to you, and you'd process it to drinking-water standards, then turn around and sell it back to them at a profit?"

In the old days, the challenge was to disperse or discharge highly treated wastewater, but over the years, the cost of treating and disposing of wastewater has dramatically increased. As it turns out, in the intervening years, reclaimed water has become a commodity. Known as recycled water, reclaimed water, purified water, or reuse, it's not just for industrial and irrigation uses anymore. Water-poor communities around the US, with ever-increasing demands, are either putting their highly treated wastewater directly into the water supply, recharging aquifers or rivers or wetlands, or considering something similar.

In Tampa Bay, the city of Clearwater spent years working on a plan to turn wastewater into drinking water. Such a proposal goes well beyond engineering, science, and resource management and into public outreach and acceptance. The cost of buying water from Pinellas County, which buys it from Tampa Bay Water, is ever-increasing. Clearwater wanted to be the master of its own fate by producing it itself and controlling the cost.

As of the first quarter of 2019, the project was on hold because of costs, but it's just a matter of time until the cost of engineering and construction is less than the price of paying someone else for water. When the dollars are right, Clearwater will likely move forward with its indirect-potable project—recycling wastewater into drinking water. In the meantime, it continues its outreach and education to the community. Public support is not a fixed point. Like water, it's fluid and must be continually renewed.

Similar pilot and demonstration projects are in place elsewhere in Florida. While Florida is subtropical, with an average annual rainfall of more than fifty inches (that's a lot), it's also mostly dependent on underground aquifers for water supply. As development continues, there is less open ground for rainfall to seep into the aquifer for recharge. Plus, the rain doesn't fall evenly, and it falls mostly in the summer. The key to water supply is reliability, and wastewater is reliable. Water suppliers know what goes out, and they know what to expect in returned resource volume.

The numbers are staggering according to Farrell. "Reuse is so plentiful among the Authority members that if they start moving into potable reuse, they wouldn't to need any more water from Tampa Bay Water for possibly fifty, sixty years."

The Florida Reuse Association spent a year developing reliable information from a wide variety of sources. Its blue-ribbon panel included environmental and political representatives as well as industry and health professionals, nonprofits, for-profits, and governments. It planned to educate and engage key audiences with facts and science. It wanted to foster political support for the necessary regulatory changes to make recycling water accessible and help remove the stigma known as the yuck factor.

Local governments won't turn that kind of resource over to another government (Tampa Bay Water) without a fight. The first skirmish occurred a couple of years ago, but in 2018, the conflict became more contentious and spilled into the media.

Tampa Bay has changed media-wise in the last twenty years. Unlike during the water wars of the 1990s, when the region could turn to either the *Tampa Tribune* or *St. Petersburg Times* for news and analysis, today there is only the *Tampa Bay Times*. In a May 2018 article, a headline made clear the newspaper's take on the new issue: *"Water Wars Starting to Simmer Again in Tampa Bay."*

It led with "A simmering feud between Tampa and the regional water authority avoided boiling over Monday, but little was resolved. At a Tampa Bay Water meeting, board members requested Tampa work with the agency on the city's plan to convert tens of millions of gallons a day of highly treated reclaimed water into drinking water. Hillsborough County has a similar program in the works."

Like other traditional media, the *Tampa Bay Times* has experienced declining revenue. That meant, despite the Pulitzers, there are fewer reporters and less consistent coverage. Away from intense, unrelenting scrutiny, Tampa Bay Water has operated openly in the shadows. Florida has some of the most liberal public-access laws in the US. But if no one is looking, no one knows what to ask for. It's also complicated. And boring.

Water is boring because it's so reliable. Yes, there were Flint, Michigan, and Camp Lejune and too many other places where water quality failed the public or threatened public health, but given the number of customers served every day, the impact is relatively small. If you're affected, it's a disaster, but you are unlikely to be affected.

In this environment, when things spill into the media, there's an information gap. Reporters working on daily news don't have the time or resources to produce nuanced narratives, and, frankly, the general reader doesn't care as long as water flows when the tap is open and it doesn't cost too much. In this context, there is tremendous pressure for member governments to make no waves. There is a sense that any disruption to peace could tip the region back into water wars. No one has the stomach or the money for that.

Consequently, the issues fester as political chicanery continues in the shadows. The hydropolitics of Tampa Bay have reached middle age and are now invisible. When issues spill into the news, the focus is more on the squabbling and less on the particulars.

The Current Issue

The city of Tampa has the largest source of treated wastewater in the region. Much of that water has poured into Tampa Bay for decades, and environmentalists credited it with helping to add a freshwater source to it. The water is higher in nutrients, which is a growing concern, but even so, the flow has a material, positive impact. The highly treated wastewater has been used for other projects over the years, including residential irrigation and industrial processes. Tampa has been the caretaker of this resource for a long time. Hillsborough County, too, has reclaimed water that could be used for other purposes. For several years it's been injecting highly treated wastewater into a coastal zone in the southern part of the county. The idea was to create a freshwater (wastewater) barrier to hold saltwater intrusion back. It has more water available and has considered partnerships with Polk County and Tampa Electric Company where reuse would be injected underground to increase the potentiometric pressure elsewhere. That increased pressure would make more groundwater available as drinking water. Both governments see value in their wastewater. Neither is willing to give up the potential income stream, though both would be interested in selling it. It seemed that Tampa Bay Water preferred a gift to a purchase. And therein lies the conflict. The "gift" would include payment in the form of credits to the provider, or trades for other facilities, but in water as in life, nothing says thank you like cash.

"There's an issue about how the use of reclaimed water interplays with contracts," said Brian Armstrong, executive director of the Southwest Florida Water Management District. "The fight there isn't over availability; it's over ownership. We are looking to invest more in indirect-potable reuse and direct-potable reuse."

Back in July 2017, Tampa Bay Water general manager Matt Jordan already knew it would be an issue. He reported this in his presentation to the District Governing Board.

"We're working with both Hillsborough County and the city of Tampa, waiting on information, talking about the things," Jordan said. "There's the notion of compensation. Should the member government providing reclaimed water be compensated? We think that's a reasonable question to ask, and so we're looking at that. So there's work to be done."

Money is a highly motivating force in water policy.

In late 2018 and early 2019, that policy disagreement spilled into the newspaper. There was a relatively new face focused on the details of the issue, and she didn't like what she saw. Darden Rice, a St. Petersburg City Council member, came to politics through environmental activism. Tampa's plan to provide millions of gallons of highly treated wastewater to Tampa Bay Water was advancing internally, but no one was asking questions. Items appeared on the agenda that didn't present a full picture, and Darden was concerned. She began to ask questions during meetings: executive-board meetings, city-council meetings, and others. Darden talked to experts and politicians. She spoke to former water warriors and water worriers, and the more she learned the more concerned she became. In her analysis, the proposed deal between the city of Tampa and Tampa

Bay Water would be the effective end of the regional agreement—the contract that ended the water wars.

When the original agreement restructuring Tampa Bay Water was signed, Tampa's primary source of drinking water was and remained the Hillsborough River. While the rest of Tampa Bay Water's member governments relied on groundwater, Tampa had the river. To participate in the regional Authority, Tampa wanted an exception to exclusivity. That meant that while it signed on to the reorganization, it would be allowed to continue to use the Hillsborough River and Tampa Bypass Canal before having to buy supplies from Tampa Bay Water. That exception was granted in January 2018—for 82 million gallons per day. If more water was available from the Hillsborough River, Tampa Bay Water could apply for a permit to use it in the regional system. Tampa's 2019 proposal sought exclusion of up to 132 million gallons per day. It's called TAP—Tampa Augmentation Project. It was a significant change to the agreement, which now states, *"Section 6.13. No Amendment, Supplement or Modification of this Agreement, including but not limited to the admission of new Member Governments or withdrawal of any Member Government, shall be binding unless executed in writing by all parties hereto."*

Tampa Bay Water, as an organization, couldn't agree to Tampa's request alone. It is not a party to the original agreement. It was the product. Instead, the six member governments—Hillsborough, Pasco, and Pinellas Counties, Tampa, St. Petersburg, and New Port Richey—all needed to agree. In writing.

It wasn't just the increase in the exclusion that was of concern. If Tampa's fifty million gallons of wastewater could be drinking water, might Tampa be allowed to self-supply that water, recycling it through the water-treatment plant to create drinking water? Or maybe it could trade some of that water for some groundwater wells?

The city of Tampa prepared an agreement for Tampa Bay Water, along with a memorandum of understanding that would codify a new relationship among the members and a way around the unanimous nine votes required to change the interlocal agreement that governs the Authority. The memorandum was presented as a "sub-agreement" that would only need six votes to pass. In this new arrangement, how would the debt be allocated? What would Tampa's plan cost the other member governments? How could it sit on a board that establishes rates if it doesn't buy the water anymore? After all, members are responsible for Tampa Bay Water's debt.

Tampa Bay Water, according to the original agreement, can't do anything without the consent of the members. In 2017, attorneys for Tampa Bay Water issued the opinion that, with limited exceptions, Tampa Bay Water is the sole provider of drinking water, that Tampa cannot treat water to supply other member governments or for itself. The only way to disperse that water for public water supply is through Tampa Bay Water. So, for nearly two years, this significant issue churned in the shadows, out of the bright lights of Florida's Sunshine Laws, except for the few insiders who were watching—like Darden Rice.

She feared that this course, with Tampa's proposed agreement, presented three dangers for the region: the division of existing debt, threats to the environment, and (potentially) the unraveling of the regional Authority.

Several former elected officials involved in the establishment or operations of Tampa Bay Water wrote an op-ed in the *Tampa Bay Times*. It was published on April 10, 2019, and said in part, *"So it is with great concern that many former leaders of Tampa Bay Water view the board's coming Monday vote on Tampa's proposed potable reuse project (TAP). If approved, Tampa will be able to fully self-supply using reclaimed water, but the unintended consequences could undermine the state's only regional water success story."*

Many believed the agreement would change the utility's business model and restore disparities that Tampa Bay Water was formed to eliminate. If approved, the agreement would allow a member to fully self-supply while still voting on matters in which it has no vested interest.

After months of delay and a bit of public scrutiny, the Tampa Bay Water board delayed a decision for another year. Even so, a new Tampa mayor found herself in the middle of the dispute. On July 19, 2019, the *Tampa Bay Times* published a story that began *"Jane Castor may have a fight on her hands over Tampa's 'toilet to tap' project. Tampa's mayor pulled a funding request related to the wastewater reuse project Thursday before Council could vote it down."*

Tampa wanted funding support for its feasibility study. Tampa Bay Water wanted access to study data and more oversight of the research, so the city withdrew its request for $1.6 million.

Attention fades, the media look elsewhere, and the faces of water leadership change, taking their corporate knowledge with them. For the people in Tampa Bay, I hope Darden Rice doesn't go with them. Still, because of her and people like her, area residents can take water for granted. Water in Tampa Bay can be eyes-roll-back boring. Nearly everyone involved, including the member governments and the public they serve, likes it that way.

"Sometimes the leading edge is the bleeding edge," said the District's executive, Armstrong, pointing out that none of this is easy, but in the end, it could be worth it.

No one saw it coming.

While the issue between Tampa and Tampa Bay Water and reclaimed water was developing and resolving and developing again, something else was happening.

This chapter was supposed to end on a hopeful note. If Tampa Bay and Tampa Bay Water were the only focus, it would be promising. However, as Tampa Bay celebrates twenty-five years of planning and investment and (mostly) collaboration, it now faces an unanticipated challenge. The newly formed Polk Regional Water Cooperative filed a placeholder application to withdraw water from the Alafia River—a source already tapped by Tampa Bay Water. The cooperative forced Tampa Bay Water to begin a real application to expand its withdrawals. At the same time, the cooperative challenged the about-to-be-issued permit of the Peace River Manasota Regional Water Supply Authority—at the farthest reaches of the Peace River downstream.

Despite the District's promise of nearly $300 million, the Polk County Regional Water Cooperative threatened the water peace in Tampa Bay and, ironically, the Peace River basin. The attorney guiding the Polk Cooperative? Ed de la Parte. Some blamed him and another recurrent player in past water issues: Gene Heath, formerly with the District and the Authority in Tampa Bay.

The newly minted Polk County Regional Water Cooperative budgeted less than $100,000 to challenge the water-use permit of the Peace River Manasota Regional Water Supply Authority. That authority is approximately one hundred miles downstream of the cooperative and fifty miles south of Polk's nearest member government. It budgeted $500,000 to defend its permit.

The cooperative said that a new fifty-year Peace River permit might someday, somehow interfere with its ability to eventually have a reservoir one hundred miles upstream, near the headwaters of the Peace River. The thing is, while a reservoir in Polk County has been on the list of possible projects for decades, it's never been near the top and would never be funded by the District. Because of the low water levels in the region, it's unlikely a reservoir is possible. The District invested nearly $300,000 in new water supplies and projects to create new supplies in Polk County, but a reservoir isn't on its list.

As part of the settlement between the cooperative and the Peace River Manasota Regional Water Supply Authority, which had been working on their permit for three years, agreed that after Polk builds a reservoir and gets a permit, the Authority, will reduce the take from the Peace River by a specific amount. It will not do so until the pumps are turned on at a Polk Reservoir. Definitely not in the foreseeable future.

In Central Florida, the Central Florida Water Initiative has rank-ordered projects to be developed—among them, the Taylor Creek Reservoir. Currently Taylor Creek is shared by the city of Cocoa with a lease from Deseret Ranch, owned by the Mormon Church. The reservoir is already in use. The St. Johns River Water Management District identified Taylor Creek as a possible regional alternative drinking-water source. Several times Orange County tried to wrest management away from the city of Cocoa. But Cocoa was not about to give up its reservoir and was prepared to walk away from regional cooperation. It took five years of negotiation with a mediator to arrive at a contract that all parties, include the city of Cocoa, Deseret Ranch, the Toho Water Authority, the city of Orlando, and Orange County would agree to. Taylor Creek is part of a larger undertaking that will include taking water from the St. Johns River. The last time a local government wanted to take less than one-half of 1 percent of the river for a water project, there was intractable opposition. The case spent several years in court, challenged by an environmental group from Northeast Florida and by the city of Jacksonville. Unlike the Peace River, the St. Johns River flows north. Ed de la Parte was the lawyer for the local government. It got its permit to take a sip of the St. Johns River.

The plan to take tens of millions of gallons from the St. Johns River will be ramped up over time. For now, it's mostly lost on the public. Too far in the future. Boring.

Beyond drinking water, water quality is at issue throughout Florida. 2018 was a disaster in terms of red algae, blue-green algae, and red tide, to name a few scourges. The impact on tourism was devastating. The effect on waterfront homeowners was awful. And the outbreaks persisted.

Residents in North Central Florida and residents in Southwest Florida are fighting the spread of phosphate strip mining. Strip mining completely disrupts and destroys the hydrogeology. Phosphate mining has been going on in Florida since phosphate rock was discovered in 1889. Time and technology have changed things. Strip-mining technology has become entirely invasive, entirely destructive, turning the ground upside down. Then, there is the occasional spill from toxic-byproduct water, the leftover from phosphate processing and the occasional sinkhole that opens under the toxic-phosphate retaining ponds. The spills typically run to the closest water body, sinkholes release into the aquifer.

The competition for water—and the need to protect it—will continue as long as demand increases and people want it for recreation. Forever. The question becomes, how will the people involved manage the politics so that everyone shares the water and the costs of developing, delivering, and protecting it—financial and otherwise?

"Right now, it's peaceful, and water is being delivered, and everything seems to be good," said Mark Farrell. "Let's hope it lasts."

Reflections on the Water Wars

On September 1, 2001, Jack Ward Thomas, former chief of the Forestry Service, was a guest on National Public Radio's *Morning Edition* program. He said that it is almost magical when a convergence of natural events and attitudes shifts a public debate. This convergence of politics, laws, economics, and public opinion is magic.

"This is a time when we can put it together and do a better job," Thomas said of the policy fight over forest management. "I believe we can do that. I believed it before, and sometimes the old magic worked, and sometimes it didn't."

Environmentalists are concerned that trees are harvested for economic gain, while government officials insist that harvesting timber is a critical part of managing the land. Clearly, these conflicts have a rhythm all their own, pulsating between the needs of the natural system and the needs of people.

In the summer of 2001, like today, fires raged across the western part of the US. There, environmentalists and foresters who argued over policy agreed that fire suppression had negatively changed the natural environment. And climate change is creating conditions for bigger, longer, hotter fires. Eventually, the two sides agreed that policy change was necessary to preserve precious forests. They made a choice to work together.

The same is true of our conflicts over water. The confluence of factors in Tampa Bay put the right people in the right place at the right time, and Mother Nature withheld rainfall. Just then, real change was possible.

We all made it through the water wars. For some, it was the highlight of their careers; for others, it was just another phase of an ongoing battle. Still, others have retired, too tired to go on, or resigned themselves to the notion that the torch must be passed. Few retired satisfied that the battle was won. Many have been lost along the way.

Who gained in the battle so fiercely fought? Clearly, the District's policy prevailed—the seawater-desalination plant and the other alternative-supply projects stand as a testament to it—no matter how fraught its beginnings or its operation. But at what cost did the District win, and was it worth it? The answer lies in whether you agreed with the policy outcome. If you did, it was worth it. If you didn't, it was not. Even those involved disagree whether the District had to go so far to push change. Some say that differences could have been resolved amicably. Others suggest that the cooperative approach had failed and would have simply perpetuated the status quo—something the District, the public, and the affected communities couldn't tolerate.

Beyond the policy outcome—or perhaps underscoring it—is how the District prevailed and what that suggests for other policy battles and the people who fight them.

Thinking back on it, the single most important thing the District did, the winning proposition, resided in the request for and focus on the support of the public. This was not lip service as so many government and consultant efforts are. This was a complete, all-out, take-no-prisoners effort to secure public support and use it as the base to launch a policy that the staff, board, and others collectively believed to be in the public interest. A time for magic.

It's important to remember that the District, like most government organizations, turned to the public only when it had nowhere else to go. The court of public opinion was the court of last resort. For whatever reason, in policy battles, this is more the exception than the rule.

In the last twenty years, the laws have called for public input—and governments (or their consultants) have complied with elaborate processes that take into consideration the public's view of processes and projects. Yes, projects are changed because of public input. But how often does a local government lead a policy-change effort? Mostly, they respond to the efforts led by others.

We need only look to South Florida and the Everglades Restoration Project to see how this works. It's a multibillion-dollar, multidecade undertaking with competing special interests and vested populations that are both vocal and engaged. There are also those communities that are voiceless and marginalized. Communication is a critical element for building the necessary relationships that make the restoration projects workable. While communications will be part of every individual project, the weight allotted to communication in the contracting process is usually a tiny fraction, though everyone says it's essential. The participants in the process agree it's essential. Yet communication is lightly weighted. Why? Part of the answer is cultural—the problem requires an engineering solution, so the decisions are made by engineers, hydrologists, geologists, and any number of other ologists—most of whom are middle-aged, male, and white. This means the approach to problems tends to be homogenous, reflecting the values of the dominant culture. That isn't to say that this group doesn't value communication; it's just not their answer to the question. Other cultures would provide other alternatives.

In Tampa Bay, real change came when it was abundantly clear that we couldn't continue as we always had. Communication created the conflict, and communication created the alternatives. The question for Tampa Bay becomes, will we continue to communicate? The answer is "It depends on the people." A quick review of chapter 18 suggests that some things have changed, some have not, and some still prefer to operate in the shadows.

There was opposition to virtually every water-supply project proposed by the resolution of the water wars, but it was narrowly focused and dispersed. There was no groundswell of opposition or support, so things continued, projects were developed, and for the most part, governments tried to make the water-supply plan work.

Studying water issues around Florida, it's not hard to see where the next battles will be. Look for places with vibrant growth and limited access to freshwater. We can identify the issues in Central and South Florida, but these are the obvious places, and there is already work underway to create collaboration. For future fights, look for places where a community with growing needs is next to one with a limited population, limited financial resources, and plenty of water. Watch for a move to modify Local Sources First, and see who's behind it. Watch for a movement to create a statewide water board or a statewide distribution system.

In August 2023, President Biden's Infrastructure Advisory Council recommended privatization and long-term leases of water systems to help manage what the council warned is a looming crisis facing the nation's aging water infrastructure. It also recommended a cabinet-level position for a water secretary. This was met with a chorus of concerns from nonprofits and watchdogs alike who pointed out that the investment banker chairing the group holds infrastructure properties in his existing portfolio. Which public? What interest?

In 2021, the BBC prepared a report regarding water shortages and water wars. It cited Iraq; Cape Town, South Africa; Lima, Peru; Rome, Italy, and more as examples. It said, "Unprecedented levels of dam building and water extraction by nations on great rivers" are occurring every day. There are people working on this, but there's not much time and as we learned in Tampa Bay, water is complicated technically and politically. With so much at stake for so many on a local, regional, national, and global scale how will leaders determine the public interest? Which public? Whose interest? If Tampa Bay has a lesson to share it's that planning and money and a pressing need to cooperate can drive solutions if we start now and together.

As leaders in Tampa Bay said repeatedly, it's hard, but it's the right thing to do.

Timeline

Date	Action
1960	Southwest Florida Water Management District established in response to flooding from hurricane Donna
1972	Water Resources Act begins to establish District authority
1975	West Coast Regional Water Supply Authority established to provide local governments the ability to cooperate in developing new water supplies.
1980-1990	A period of cooperation between the District and the Authority in which new water supplies were developed.
	Cross Bar wellfield begins operations
	Cross Bar wellfield increases production
1988	District begins Water Resource Assessment Project
1992	St. Petersburg applies for renewal of its public supply water use permit
Jan.1994	Steve Monsees speaks to the District Governing board
Feb. 1994	District holds special meeting to determine the water necessary to meet "public health and safety"
Mar. 1994	Local governments, the District, environmentalists, and others choose sides
May 1994	Driest May in 100 years
July 1994	Governing board adopts an emergency order effectively capping the amount of water that can be pumped for public supply in Tampa Bay
1995	Litigation and negotiation continue
1996	Pinellas County sues water activists
1996	St. Petersburg's permit application denied
1997	Administrative Hearing Order issued, everyone declares victory, but no one wins. Negotiations commence on a new approach to water supply development
1998	Authority reorganized into Tampa Bay Water
1999-present	New activists, new board members, new water supply problems

Story Participants and Roles

	1970s	1980s	1990s	2020
Name	Title			
Honey Rand			District Staff	Consultant
Pete Hubbell	District Staff	District Staff	District Exec	Consultant
Sonny Vergara	District Staff	St. Johns Exec	WSA Gen. Mgr	District Exec
Gene Heath		Authority	Consultant	District Exec
Roy Harrell		Gov. Bd.	Gov. Bd.	
Doug Manson		Lawyer/Ag.	Lawyer/St. Pete	Lawyer/Tampa
Steve Monsees			Activist	
Gene Schiller			District Exec	District Exec
Mark Farrell		District Staff	District Exec	Consultant
Bill Johnson	Hills Utility	St. Pete Utility	St. Pete Utility	St. Pete Utility
John Parker		District Staff	District Staff	District Staff
Steve Seibert			Pinellas Comm	Secretary DCA
Dave Fischer			Mayor St. Pete	Mayor St. Pete
Ed de la Parte	Lawyer Authority	Lawyer Authority	Lawyer/Pinellas	Lawyer/Pinellas
Joe Davis			Gov. Bd.	
Ed Collins			Pasco Comm	Consultant
Gilliam Clarke			Activist	Activist
Eilene Hart	Activist	Activist	Activist	Activist
Judy Williams			Activist	Activist
Chuck Rainey	Pinellas Comm			
Derrill McAteer	District Chair			
Jerry Maxwell			Authority Gen. Mgr.	Authority Gen. Mgr.
Pick Talley	Pinellas Utility	Hills Comm	Pinellas Utility	Pinellas Utility
Dick Greco	Mayor Tampa		Mayor Tampa	Mayor Tampa

References

Beeman, K. (1994, June 11). Water cutbacks may mean shorter shower, fewer flushes. *Tampa Tribune*, B1.

Beeman, K. (1994, June 4). More bay area wells may turn off the taps. *Tampa Tribune*, B1.

Blake, N. M. (1980). *Land into water—Water into Land*. University Presses of Florida.

Commissioner suggests group. (1994, June 30). *St. Petersburg Times*, B2.

Davis, A. (1999, May 14). Pinellas officials to make apology. *St. Petersburg Times*, B4.

de la Parte, E. (1994, July 10). Times plays cheerleader to district power grab. *St. Petersburg Times*, D3.

Derr, M. (1989). *Some kind of paradise: A chronicle of man and the land in Florida*. William Morrow and Company.

Fernald, E. A. (1984). *Water resources atlas of Florida*. Florida State University.

Florida Water Resources Study Commission. (1956, December). Florida's water resources: A study of the physical, administrative, and legal aspects of water problems and water management. Florida Water Resources Study Commission.

Heller, J. (1996, August 13). Document clouds findings on water. *St. Petersburg Times*, B1.

Heller, J. (1996, June 25). Some people will do anything to make a point. *St. Petersburg Times*, B1.

Heller, J. (1996, September 21). Swiftmud, lawyer argue over paralegal's actions. *St Petersburg Times*, B1.

Heller, J. (1997, September 15). Water pumping limits threaten efforts for truce. *St. Petersburg Times*, B1, B3.

Heller, J. (1999, September 24). Water utilities general manager clashes with counsel. *St. Petersburg Times*, A1.

Hennessey, L. (1980, February 1). Letter to interoffice correspondence.

Houk, K. (1995, January). *An assessment of public attitudes on water issues.*

Hubbell, P. (1996, July 28). Executive director, Southwest Florida Water Management District. *Tampa Tribune,* D2.

Khator, R. (1996, July). *Water issues survey.* University of South Florida.

Klages, W. J. (1995, December 14). *Focus group research: Water issues.* Evans-Klages.

Landry, S. (1994, June 15). Area's water bill may be coming due. *St. Petersburg Times,* B1.

Landry, S. (1994, June 23). Accord in short supply as agencies meet. *St. Petersburg Times,* B1.

Maloney, F. E. (1968). *Water law and administration: The Florida experience.* University of Florida Press.

Manai, L. (1995, May 5). Water trip a "successful failure." *St. Petersburg Times,* B1.

McLean, R. (1994, June 28). The big lie. *St. Petersburg Times,* D3.

Milne, S. (2021, August 16). How water shortage are brewing wars. BBC. https://www.bbc.com/future/article/20210816-how-water-shortages-are-brewing-wars

National Oceanic and Atmospheric Administration. *50 years of population change along the nation's coasts: 1960–2010.* http://seaserver.nos.noaa.gov/projects/population/population.html.

Nitnaware, H. (2023, July 7). Over 50% of world population to face water shortages by 2050: FAO urges countries to prepare for it. https://www.downtoearth.org.in/news/water/over-50-of-world-population-to-face-water-shortages-by-2050-fao-urges-countries-to-prepare-for-it-90468

Olinger, D. (1995, February 23). Water wars may soak up millions. *St. Petersburg Times,* B1, B3.

Pedreira, D. (1996, January 19). Pinellas lawsuit gets hot response. *Tampa Tribune,* B1.

Pedreira, D. (1999, May 13). Pinellas settles activist lawsuit. *Tampa Tribune,* A1.

Penney, V., & Muyskens, J. (2023, August 16). Here's where water is running out in the world—and why. *Washington Post*. https://www.washingtonpost.com/climate-environment/interactive/2023/water-scarcity-map-solutions/

Pettegrew, L. (1995, May). *Public response to alternative water resources*. Southwest Florida Water Management District.

Pinellas County Board of County Commissioners. (1996, January 16). BOCC Workshop. , Pinellas County Public Records.

Pinellas County Board of County Commissioners. (1996, January 23). Regular Meeting.

Pinellas County Public Information Department. (1992). Request for Proposals, Pinellas County.

Pskowski, M. (2023, June 5). Without access to water lines, Texas colonias residents are pulling water from the desert air. *Texas Tribune*. https://www.texastribune.org/2023/06/05/texas-colonias-water-hydro-panels-hueco-tanks-el-paso/

Rand, H. (1994, June 21). Fact sheet. Personal files.

Rojanasakul, M., Flavelle, C., Migliozzi, B., & Murray, E. (2023, August 28). America is using up its groundwater like there's no tomorrow. *New York Times*. https://www.nytimes.com/interactive/2023/08/28/climate/groundwater-drying-climate-change.html?campaign_id=9&emc=edit_nn_20230829&instance_id=101398&nl=the-morning&regi_id=72010475&segment_id=143147&te=1&user_id=3ebc769e129c0b70daef603b22a98569

Seaton, N. (1995, March 21). Rainey wants city to toe line in water fight. *St. Petersburg Times*, B2–B3.

Seaton, N. (1996, January 17). Pinellas going on offense in water wars. *St. Petersburg Times*, B1.

Seaton, N. (1996, January 22). Water ad campaign goes against flow. *St. Petersburg Times*, B1, B3.

Seaton, N. (1998, January 18). Ads will pump water words out. *St. Petersburg Times*, B1, B4.

Seibert, S. (1997, May 15). Letter to Pinellas County Commission.

Seibert, S. (1997, August 8). Letter to Pinellas County Commission.

Southwest Florida Water Management District. (1926, January 26). Governing board meeting. Southwest Florida Water Management District.

Southwest Florida Water Management District. (1996, December 20). Governing board meeting. Southwest Florida Water Management District.

Southwest Florida Water Management District. (1997, September 8). Governing board meeting. Southwest Florida Water Management District.

Southwest Florida Water Management District. (1997, September 23). Governing board meeting. Southwest Florida Water Management District.

Southwest Florida Water Management District. (1998). Mission statement. Southwest Florida Water Management District.

St. Petersburg City Council. (1994, February 7). Resolution. St. Petersburg City Council.

Tebeau, C. W. (1980). *A history of Florida.* University of Miami Press.

Texas A&M University. (2023, August 6). Analysis of over 7,000 reservoirs shows worldwide water reserves are depleting. *SciTechDaily.* https://scitechdaily.com/analysis-of-over-7000-reservoirs-shows-worldwide-water-reserves-are-depleting/

United Nations Department of Global Communication. (2023, February 13). United Nations Department of Economic and Social Affairs. https://www.un.org/en/desa/un-assembly-president-urges-governments-send-top-leaders-water-conference

U.S. Geological Survey (1968). *Hydrologic effects of groundwater pumping in the Florida Aquifer in northwest Hillsborough, northeast Pinellas and southwest Pasco Counties, Florida.* U.S. Geological Survey.

Water Management District Review Commission. (1995, December 29). *Bridge over troubled water.* Water Management District Review Commission.

Water power. (1973, September 16). *St. Petersburg Times,* D2.

Water shortage real. (1994, April 16). *Tampa Tribune,* D2.

Index

www.ingramcontent.com/pod-product-compliance
Lightning Source LLC
LaVergne TN
LVHW101505200926
848483LV00001B/233